Vol. 36. **Emission Spectrochemical Analysis.** By Morris Slavin
Vol. 37. **Analytical Chemistry of Phosphorus Compounds.** Edited by M. Halmann
Vol. 38. **Luminescence Spectrometry in Analytical Chemistry.** By J. D. Winefordner, S. G. Schulman, and T. C. O'Haver
Vol. 39. **Activation Analysis with Neutron Generators.** By Sam S. Nargolwalla and Edwin P. Przybylowicz
Vol. 40. **Determination of Gaseous Elements in Metals.** Edited by Lynn L. Lewis, Laben M. Melnick, and Ben D. Holt
Vol. 41. **Analysis of Silicones.** Edited by A. Lee Smith
Vol. 42. **Foundations of Ultracentrifugal Analysis.** By H. Fujita
Vol. 43. **Chemical Infrared Fourier Transform Spectroscopy.** By Peter R. Griffiths
Vol. 44. **Microscale Manipulations in Chemistry.** By T. S. Ma and V. Horak
Vol. 45. **Thermometric Titrations.** By J. Barthel

Thermometric Titrations

CHEMICAL ANALYSIS

A SERIES OF MONOGRAPHS ON ANALYTICAL CHEMISTRY AND ITS APPLICATIONS

Editors

P. J. ELVING · J. D. WINEFORDNER

Editor Emeritus: **I. M. KOLTHOFF**

Advisory Board

J. Badoz-Lambling	J. J. Lingane	Charles N. Reilley
George E. Boyd	F. W. McLafferty	E. B. Sandell
Raymond E. Dessy	John A. Maxwell	Eugene Sawicki
Leslie S. Ettre	Louis Meites	Carl A. Streuli
Dale J. Fisher	John Mitchell	Donald E. Smith
Barry L. Karger	George H. Morrison	Wesley Wendlandt

VOLUME 45

A WILEY-INTERSCIENCE PUBLICATION

JOHN WILEY & SONS
New York / London / Sydney / Toronto

Thermometric Titrations

J. BARTHEL

Department of Chemistry
University of Regensburg
Regensburg, Germany

with a chapter on

Instrumentation in Titration Calorimetry

by R. WACHTER
Department of Chemistry
University of Regensburg, Germany

A WILEY-INTERSCIENCE PUBLICATION

JOHN WILEY & SONS
New York / London / Sydney / Toronto

Copyright © 1975 by John Wiley & Sons, Inc.

All rights reserved. Published simultaneously in Canada.

No part of this book may be reproduced by any means, nor transmitted, nor translated into a machine language without the written permission of the publisher.

Library of Congress Cataloging in Publication Data:

Barthel, Josef, 1929–
  Thermometric titrations.

  (Chemical analysis ; v. 45)
  "A Wiley-Interscience publication."
  Translated from the German.
  Bibliography: p.
  1. Volumetric analysis. 2. Calorimeters and calorimetry. I. Title. II. Series.
QD111.B273   545'.2   75-17503
ISBN 0-471-05448-8

Printed in the United States of America

10 9 8 7 6 5 4 3 2 1

PREFACE

Thermometric titrations encompass a multitude of methods, all of which have a common basis in calorimetric techniques. The possible applications of thermometric titration range from analytical thermometric and enthalpimetric methods to precision calorimetric methods for the determination of thermodynamic data in single- and multistep equilibrium processes. This book is intended to cover the range of these applications, to present their theoretical basis in an easily understandable manner, and to acquaint the reader with the experimental methods and instrumentation, particularly with the intrinsic possibilities of the thermometric method itself. The wide-ranging analytical applications are discussed in detail, in a way which shows that these applications by no means exhaust the possibilities of the thermometric method.

In preparing this book I received kind assistance from many sides. The manuscript was prepared in cooperation with Professor Thomas Finkenstaedt, University of Augsburg, and Professor Michael R. F. Ashworth, University of Saarbrücken, to whom I extend my thanks for their invaluable efforts to avoid losses in the process of translation from the original German. From the beginning the analytical investigations were carried out with the assistance of Mrs. Klothilde Wachter-Lenz, whom I should also like to thank for her help in preparing the bibliography. I thank Professor Rudolf Wachter, University of Regensburg, for developing the calorimetric instrumentation and writing Chapter 9. Thanks are also due to Miss Marlene Westermeier for her careful typing of the manuscript. Finally, I thank all copyright holders for their kind permission to reprint tabular material and figures.

Regensburg, Germany   J. BARTHEL
August 1975

CONTENTS

CHAPTER 1 GENERAL INTRODUCTION

1.1 Thermometric titration — 1

1.2 Thermodynamic principles — 5

    1.2.1 *Enthalpy of reaction* — 6
    1.2.2 *Entropy of reaction and Gibbs energy of reaction* — 7
    1.2.3 *Chemical potential and activity functions* — 8
    1.2.4 *Reference state and standard conditions* — 10
    1.2.5 *Chemical equilibrium* — 12
    1.2.6 *Practical application of thermodynamic principles in thermometric titration* — 13

1.3 Electrochemical principles — 14

    1.3.1 *Ionophoric and ionogenic electrolytes* — 14
    1.3.2 *Activity coefficient* — 15
    1.3.3 *Aqueous electrolyte solutions* — 17

1.4 Titrations in nonaqueous solvents — 19

CHAPTER 2 THEORETICAL BASIS FOR AN ANALYSIS OF THERMOGRAMS

2.1 Thermograms — 24

2.2 The basic equation of thermometric titration — 29

    2.2.1 *The overall thermal power* — 29
    2.2.2 *Thermal power of the mixing process* — 30
    2.2.3 *Thermal power of the reaction* — 31
    2.2.4 *Thermal powers caused by the calorimeter* — 33
    2.2.5 *The basic equation of thermometric titration* — 33

2.3 Evaluation of titration diagrams — 35

    2.3.1 *Mathematical description of thermograms of titrations with incomplete titration reaction* — 35

2.3.2	Mathematical description of thermograms of titrations with complete titration reaction	37
2.3.3	Calculation of the end point and an approximate enthalpy of reaction	38
2.3.4	The section method	43
2.3.5	The tangent method	46
2.3.6	Simultaneous determination of ΔG_R°, ΔH_R°, and ΔS_R° from a thermogram	47
2.3.7	Other methods of evaluation	49
2.3.8	Analysis of binary mixtures and multistep reactions	53

CHAPTER 3 METHODS OF THERMOMETRIC TITRATION

3.1 The TET method — 56

3.1.1	Description of the method	56
3.1.2	Determination of heat capacity C, specific heat c_B, and cooling constant κ	58
3.1.3	Continuous or discontinuous addition of titrant	61
3.1.4	Recent developments of TET technology	63
3.1.5	Thermometric titration as a method for determining enthalpy of mixing	64

3.2 The DIE method — 66

3.2.1	Description of the method	66
3.2.2	Theoretical basis of the method	68
3.2.3	Applications of the method	69

3.3 Continuous-flow enthalpimetry — 70

3.3.1	Description of the method	70
3.3.2	Theoretical basis of the method	70
3.3.3	Application of the method	71
3.3.4	Flow microcalorimetry	72

3.4 Special methods for increasing the accuracy of titration — 73

3.4.1	Combination with a secondary reaction of high heat change	73
3.4.2	Indirect thermometric analysis	74
3.4.3	Kinetic titration	74

3.5	**Methods with amplified end-point indication**	**75**
	3.5.1 *Catalytic thermometric titration*	75
	3.5.2 *Intensification of end point through high heat of dilution*	75

CHAPTER 4 ACID-BASE TITRATIONS IN AQUEOUS SOLUTIONS

4.1	**The heat effect of standard reactions**	**77**
	4.1.1 *Calorimetric determination of the heat of neutralization*	77
	4.1.2 *Electrochemical determination of the heat of neutralization*	80
	4.1.3 *Protonization of Trishydroxymethylaminomethane as a new standard reaction*	81
4.2	**Titration of strong acids with strong bases in aqueous solutions**	**81**
4.3	**Titration of weak acids or weak bases in aqueous solutions**	**82**
	4.3.1 *Weak monoacids and monobases*	82
	4.3.2 *Mixtures of monoacids and monobases*	85
	4.3.3 *Polyvalent acids*	89
	4.3.4 *Displacement of a function*	94
	4.3.5 *Formation and decomposition of polyanions*	95

CHAPTER 5 TITRATIONS IN NONAQUEOUS SOLVENTS

5.1	**Determination of the purity of solvents**	**98**
5.2	**Lewis acids and Lewis bases**	**99**
5.3	**Organometallic compounds**	**101**
5.4	**Brönsted acids and Brönsted bases**	**102**
5.5	**Special end point indication techniques**	**104**
	5.5.1 *Catalytic thermometric titration*	104
	5.5.2 *End-point indication with the help of a large heat of dilution*	107
5.6	**Further titrations in nonaqueous solvents**	**108**
	5.6.1 *Unsaturated compounds*	108
	5.6.2 *Acetylation*	108

5.7	Thermometric titration in molten salts	109

CHAPTER 6 THERMOMETRIC PRECIPITATION TITRATION

6.1	Theoretical basis of precipitation titration	113
6.2	Special types of precipitation reaction	118
	6.2.1 *Halides and cyanides*	118
	6.2.2 *Fluorocomplex ions*	119
	6.2.3 *Hydroxides and basic salts*	120
	6.2.4 *Tetraphenylborates*	120
	6.2.5 *Perchlorates*	122
	6.2.6 *Oxalates*	122
	6.2.7 *Phosphates*	123
	6.2.8 *Carbonates*	123
	6.2.9 *Sulfates*	123
	6.2.10 *Catalytic thermometric titrations*	124

CHAPTER 7 THERMOMETRIC REDOX TITRATIONS

7.1	Potassium dichromate titrations	127
7.2	Titrations with permanganate	130
7.3	Titrations with metal ions	131
7.4	Hydrogen peroxide titrations	132
7.5	Titrations with arsenious acid, iodine, hypochlorite, and further agents	133

CHAPTER 8 FORMATION OF COMPOUNDS AND COMPLEXES. PROBLEMS OF COMPLEXOMETRIC TITRATION

8.1	Attempts to determine the composition of complex compounds	135
8.2	Determination of thermodynamic data of a complex-formation reaction	138
	8.2.1 *The method of continuous variation*	138

8.2.2	*Other methods for the approximate determination of the constant of formation K_c*	**139**
8.2.3	*Entropy titration*	**140**

8.3 Complexometric titrations — **149**

8.3.1	*Complex-ion formation*	**149**
8.3.2	*Chelate formation*	**149**
8.3.3	*Complexes of aminopolycarboxylic acids*	**150**

CHAPTER 9 INSTRUMENTATION IN TITRATION CALORIMETRY BY R. WACHTER

9.1 Principles of design — **159**

9.1.1	*Heat exchange between the calorimeter and its environment*	**159**
9.1.2	*Temperature measurement in titration calorimetry*	**167**
9.1.3	*The addition of titrant solution*	**176**
9.1.4	*Calibration of titration calorimeters*	**177**

9.2 Calorimetric equipment for special purposes — **181**

9.2.1	*Titration calorimeters for accurate enthalpy change measurement*	**181**
9.2.2	*Titration calorimeters for predominently analytical work*	**191**
9.2.3	*Microcalorimeters*	**196**

INDEX — **201**

Thermometric Titrations

CHAPTER

1

GENERAL INTRODUCTION

1.1 THERMOMETRIC TITRATION

Thermometric titration in the narrow, original sense is a volumetric method in which the heat effect of the titration reaction is used to measure the titer of a sample. It is a universally applicable analytical method, because every chemical reaction

$$x_1A_1 + x_2A_2 + \cdots + x_nA_n \rightleftarrows z_1P_1 + z_2P_2 + \cdots + z_nP_n \qquad (1.1)$$

entails a heat of reaction Q_R. Its precision is that of volumetric methods in general; nowadays, standard equipment allows an accuracy of 0.2 to 0.5% and precision equipment, of 0.1 to 0.2%. In (1.1), symbol A_i refers to the initial reactants and P_i to the end products, and x_i and z_i refer to the relevant stoichiometric coefficients.

The heat evolved by reaction (1.1) can be used in two ways in analytical methods:

1. For determining an end point of titration, when the temperature of the sample is measured during the titration process as a function of the volume of titrand added [see thermometric enthalpy titration (TET) techniques, Section 3.1].
2. For determining the unknown amount of titrand, when the measured total heat of the titration process is compared with the known molar heat of this reaction or any other amount of heat that is related with the titration reaction in a calibration process [see direct injection enthalpimetry (DIE) techniques, Section 3.2].

The expression *thermometric titration* was probably first used by Dean and Watts.[1] Dutoit and Grobet,[2] the founders of this method, speak of "une nouvelle méthode de volumétrie physico-chimique." Ewing[3] uses the term *thermal titration*; Jordan et al., *enthalpy titration*[4,5] and *thermometric enthalpy titration* (TET)[6]; and Bjerrum,[7] *calorimetric titration*. In 1957 the Thermoanalytical Titrimetry Symposium of the American Chemical Society voted for the term *thermometric titration*,[8] but this term has not been generally accepted yet.

Whereas Dutoit and Grobet stated that a thermometric titration need not take more than 1 hr, the present state of technology enables it to be performed as fast as other automatic methods of titration.

Titration based on the heat of reaction renders the thermometric methods universally applicable; acid-base titrations (Chapter 4), precipitation titrations (Chapter 6), redox titrations (Chapter 7) and complexometric titrations (Chapter 8) are all equally feasible if the observed rise of temperature is sufficiently large. Technological developments (Chapter 9) in the field of titration calorimeters have reached a stage in which titration in the micromole range is possible. In addition special titration procedures have been developed (Chapter 3); in catalytic thermometric titration, for example, end-point indication is even independent of the heat of the titration reaction. A special advantage of thermometric titration is that it is largely independent of special features of the sample, such as ionic strength or solvent, if these do not influence the titration reaction as such. Colored or colloidal solutions or slurries can be handled. In contrast to the electrodes of electrochemical methods, the temperature sensor as measuring instrument is inert and does not falsify results through reaction with components of the sample.

Applications in a wide field have accompanied the developments of thermometric titration from the very first. Bell and Cowell[9] used an early TET technique in agricultural chemistry, and it seems that, in 1886, C. B. Howard was already using a DIE technique for an estimation of the strength of sulfuric acid;[10] early applications to technical organic chemistry are due to Somiya, for example, the determination of acetyl value of oils and fats[11] and iodine value of unsaturated fats.[12] Today's applications concern analytical work in metallurgy (see Section 3.2.3); coal,[13] tar, and petroleum products (Chapter 5); cements and industrial silicates (see Section 3.2.3); paper manufacture;[14] surfactants;[15,16] process control for various products (see Chapters 3 and 9); the investigation of biological relevant products (see Sections 3.3.4 and 8.2); and the analysis of pharmaceutical products[17–20] and products used in clinical chemistry.[21,22]

A specific and characteristic aspect of thermometric titration distinguishes it from other methods: As "enthalpy titration" it can be contrasted to "Gibbs energy titration," a representative example of which is potentiometry. Titration methods with visual end-point detection by the color change of an indicator are also Gibbs energy titrations. In an enthalpy titration the change of enthalpy with addition of the titrant is followed; in the Gibbs energy titration it is the change of the Gibbs energy. Jordan has repeatedly stressed this fact, which justifies the name "enthalpy titration" for the thermometric method.[4,6,8,23] The thermodynamic property

$$\Delta H = \Delta G + T \Delta S \tag{1.2}$$

(where H is enthalpy, G Gibbs energy, S entropy, and T temperature in degrees Kelvin) illustrates the difference. The entropy change during the reaction may in some cases permit the end-point indication on the basis of a changing enthalpy, whereas it is impossible on the basis of Gibbs energy. Thus the titration of boric acid is impossible in a ΔG titration,[24] whereas a ΔH titration gives a sharp end point (Fig. 1.1).

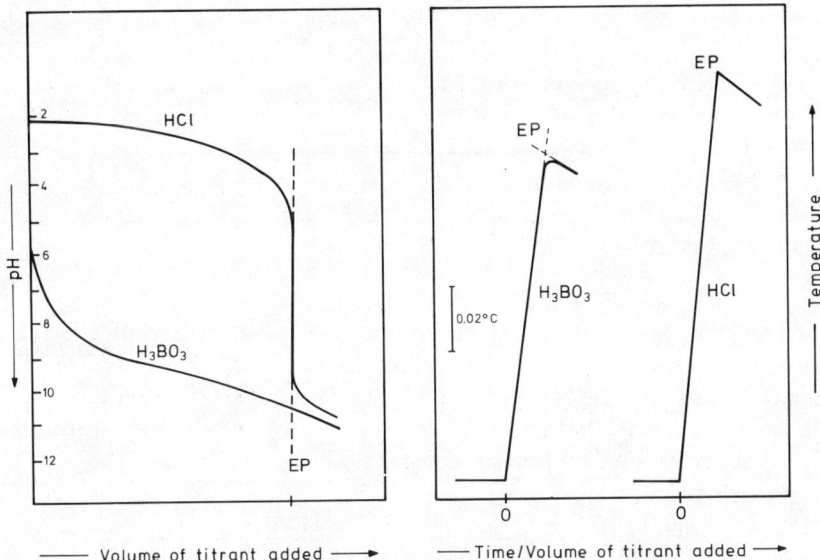

Fig. 1.1. Comparison of Gibbs energy titration (potentiometric) and enthalpy titration (thermometric).[23] Curves: 0.01 M acids titrated with standard NaOH in aqueous solutions at 25°C. EP: end point.

This aspect of thermometric titration is also seen in the complexometric titration of Ca^{2+} alongside Mg^{2+}, using EDTA.[4] The Gibbs energy of reaction is -15 kcal mol^{-1} for the reaction of EDTA with Ca^{2+} and -12.4 kcal mol^{-1} with Mg^{2+}, whereas the respective reaction enthalpies are -6.5 kcal mol^{-1} and $+5.5$ kcal mol^{-1}. This means that there are only slightly differing ΔG values for the two reactions, one of which is exothermic and the other endothermic, whereas the difference in the ΔH values is 12 kcal mol^{-1}. The usual complexometric titration using eriochrome black as indicator gives only the sum of the molarities of Ca^{2+} and Mg^{2+}, but the thermometric method allows the titration of each ion in the presence of the other (Fig. 1.2).

According to the remarks just made, reaction enthalpy appears as the

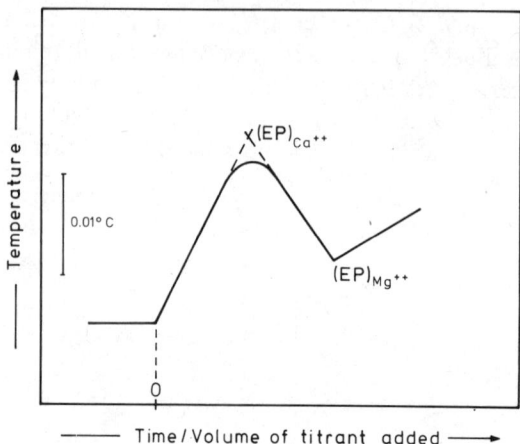

Fig. 1.2. Titration of a mixture of Ca^{2+} and Mg^{2+} with ethylenediaminetetraacetate in aqueous solution[23] at 25°C.

only relevant factor for an understanding of thermometric titration, because we have hitherto implied that the titration reaction is complete. This means that in (1.1) the reaction equilibrium is situated so far on the side of the end products P_i that no initial reactants A_i can be detected at the end of the titration. The equilibrium constant

$$K_a^{(c)} = \frac{[P_1]^{z_1}[P_2]^{z_2}\cdots[P_n]^{z_n}}{[A_1]^{x_1}[A_2]^{x_2}\cdots[A_n]^{x_n}} \frac{y_{P_1}^{z_1}y_{P_2}^{z_2}\cdots y_{P_n}^{z_n}}{y_{A_1}^{x_1}y_{A_2}^{x_2}\cdots y_{A_n}^{x_n}} \quad (1.3)$$

(where $K_a^{(c)}$ is the thermodynamic equilibrium constant, $[A_i]$ the concentration of compound A_i, and y_i the relevant activity coefficient) is then not useful for the problem. In such cases $K_a^{(c)}$ exceeds 10^{10}.

If, however, the equilibrium of the reaction is such that definite amounts of A_i are always coexistent with P_i, we observe at the stoichiometrically calculated equivalence point a more and more pronounced curvature as the equilibrium constant decreases, and thus lower accuracy and precision in titration. In the range of low equilibrium constants $[K_a^{(c)} \sim 10^{-3}$ or $10^{-2}]$, the determination of an end point becomes impossible. However, as analytical interest in the problem diminishes, thermodynamic interest increases. The method of thermometric titration has proved to be an efficient instrument for determination of thermodynamic functions ($\Delta H°$, $\Delta S°$, $\Delta G°$, or K_a) of equilibrium reactions and it has been increasingly used for this purpose since the early sixties, independent of its original and intrinsic interest. As a further new development in its thermodynamic application, a thermometric technique for the determination of heats of mixing has been proposed (see Section 3.1.3).

The kinetic aspects of titration must still be discussed in this general survey. Some particular kinetic situations can be utilized:

1. The possibility of successively titrating two components in a mixture when their rates of reaction with the titrant differ considerably.[25]
2. Benefit from a subsequent reaction of the titrant with the solvent or with an indicator component added to the reaction medium (see Sections 5.5.1 and 8.3.3) in catalytic thermometric titration.
3. Differential titration of two systems, one with and one without added catalyst, as described in Sajó's kinetic titration (see Section 3.4.3).

Apart from these methods, there is a general kinetic effect encountered in every titration procedure, whether thermometric or not. This effect results from the competition between the rates of addition of titrant and of reaction (see also Section 2.3.3).[26,27] It can lead to markedly incorrect end-point indication, especially when the titrant is run in continuously. This error can generally be eliminated in discontinuous titrant addition by the judicious choice of intervals between additions (Section 3.1.3). Chemical kinetics makes use of this effect to determine rate constants of chemical reactions (see flow methods[28]).

The interest in thermometric titration in all its various forms and applications is mirrored in many surveys of the field[3,5,6,8,23,27,29-45] and the monographs by Tyrrell and Beezer,[46] Bark and Bark,[47] and Vaughan.[48]

This introductory section has revealed a number of problems of a thermodynamic, electrochemical, and general chemical nature connected with thermometric titration; these will be dealt with in the following sections of the first chapter before the thermometric method is described in detail.

1.2 THERMODYNAMIC PRINCIPLES

In presenting the thermodynamic basis of thermometric titration, problems connected with the addition of the titrant are at first put aside. Thus the equation of the heat balance does not contain terms relating to the heat associated with the mixing of the reactants if these have different temperatures, nor is the heat of mutual dilution of the components during the process of mixing taken into account. The heat exchange between the titration vessel and its environment, as well as the heat of stirring, is also left out for the time being. If we assume that the reaction rate is sufficiently fast, there are two central problems to be discussed as fundamental for thermometric titration: the heat of reaction and chemical equilibrium.

In an adiabatically closed reaction vessel, the mixed initial reactants $A_1, A_2, \ldots, A_n$ of the titration reaction are present at the time $t = 0$ (initial state). The system is transformed into its final state, in which the end products $P_1, P_2, \ldots, P_n$ are coexistent with $A_1, A_2, \ldots, A_n$ according to the equilibrium of reaction (1.1). The temperature rise ΔT is measured with the help of a temperature sensor.

1.2.1 ENTHALPY OF REACTION

The calorimeter vessel of known content has a heat capacity C:

$$C = C_{\text{calorimeter}} + C_{\text{content}} \tag{1.4}$$

This capacity is the sum of the heat capacities of the solution in the calorimeter (C_{content}) and of the calorimeter itself ($C_{\text{calorimeter}}$). The latter includes the heat capacity of the vessel itself and the amount due to the different parts of the calorimeter (the stirrer, temperature sensor, heater, and tip of the burette) that are immersed in the reaction solution and are heated with it. Methods for measuring the heat capacity are discussed in Chapter 3.

The temperature rise ΔT, indicated by the temperature sensor, is related to the heat Q_{tot} developed in the reaction vessel according to

$$Q_{\text{tot}} = C \Delta T \tag{1.5}$$

In accordance with the simplified model, this amount of heat is produced only by the chemical reaction (1.1) taking place in the calorimeter. In Chapter 2 a more sophisticated and exact heat balance is presented; the present simplifications do not, however, impair the general validity of the statements about the thermodynamic principles involved.

In a system of heat capacity C a temperature increase of dT provokes a change dU in the internal energy U. Introducing enthalpy $H = U + pV$ gives, at constant pressure p,

$$(dH)_p = dQ_P \tag{1.6}$$

As thermometric titrations are always carried out at constant pressure, the index p that denotes it will be henceforth omitted.

The heat-producing process may be a homogeneous or heterogeneous reaction, the simplest case for a thermodynamic treatment being a reaction in the gas phase,

$$\verb|\|C\!=\!C\verb|/| + H_2 \rightleftharpoons H\!-\!\overset{|}{C}\!-\!\overset{|}{C}\!-\!H \tag{1.7}$$

Thermometric titrations in the gas phase have not been performed so far, but reactions involving a gaseous reactant, for example, a reaction according to (1.7) with the olefin dissolved in hexane,[49] or the reaction[50]

$$2OH^-(aq) + CO_2(g) \rightleftharpoons CO_3^{2-}(aq) + H_2O(l) \quad (1.8)$$

have been carried out. The usual types of titration reaction are those with a titrant in solution,

$$H^+(aq) + OH^-(aq) \rightleftharpoons H_2O(l) \quad (1.9)$$

including those with a nonelectrolyte in solution, such as the titration of a Lewis acid with dioxane in benzene as solvent,[51]

$$SnCl_4 + C_4H_8O_2 \rightleftharpoons SnCl_4 \cdot C_4H_8O_2 \quad (1.10)$$

As a thermodynamic property, the enthalpy of a system is additively made up of the enthalpies of its individual components. For a chemical reaction of the components according to reaction (1.1), the reaction enthalpy is

$$\Delta H_R = [z_1 H(P_1) + z_2 H(P_2) + \cdots + z_n H(P_n)]$$
$$- [x_1 H(A_1) + x_2 H(A_2) + \cdots + x_n H(A_n)] \quad (1.11)$$

$H(i) = \overline{H}_i$ is the partial molar enthalpy of the component i and is considered to be constant for each component during the reaction.

1.2.2 ENTROPY OF REACTION AND GIBBS ENERGY OF REACTION

Statements about the chemical equilibrium of a reaction must be based on the second law of thermodynamics. To this end the function

$$G = H - TS \quad (1.12)$$

is used. The entropy S and the Gibbs energy G are thermodynamic properties so that the following relationships are valid for chemical reactions of (1.1):

$$\Delta S_R = [z_1 S(P_1) + z_2 S(P_2) + \cdots + z_n S(P_n)]$$
$$- [x_1 S(A_1) + x_2 S(A_2) + \cdots + x_n S(A_n)] \quad (1.13)$$

and

$$\Delta G_R = [z_1 G(P_1) + z_2 G(P_2) + \cdots + z_n G(P_n)]$$
$$- [x_1 G(A_1) + x_2 G(A_2) + \cdots + x_n G(A_n)] \quad (1.14)$$

in analogy to the enthalpy balance equation (1.11) and using partial molar quantities, namely, molar entropy $S(i) = \bar{S}_i$ and molar Gibbs energy $G(i) = \bar{G}_i$ of the component i. The relation (1.12) applies for each component:

$$\bar{G}_i = \bar{H}_i - T\bar{S}_i \tag{1.15}$$

1.2.3 CHEMICAL POTENTIAL AND ACTIVITY FUNCTIONS

Designating any of the partial molar quantities $\bar{H}_i$, $\bar{S}_i$, and $\bar{G}_i$ as $\bar{X}_i$, we find that the quantities can be unified as

$$\bar{X}_i = \left(\frac{\partial X}{\partial n_i}\right)_{p, T, n_{j \neq i}}$$

where X is the thermodynamic property, H, S, or G, of the total system and n_i the amount of substance (number of moles) of the component i.

Any relationship that is valid for the thermodynamic properties of the total system holds for the corresponding partial molar quantities, for example,

$$\left(\frac{\partial [\bar{G}_i/T]}{\partial T}\right)_{p, n_i} = -\frac{\bar{H}_i}{T^2} \quad \left(\frac{\partial \bar{G}_i}{\partial p}\right)_{T, n_i} = \bar{V}_i \quad \left(\frac{\partial \bar{G}_i}{\partial T}\right)_{p, n_i} = -\bar{S}_i \tag{1.16a,b,c}$$

Another example is given by (1.12) and (1.15).

Together with the partial molar quantities $\bar{X}_i$ the chemical potential μ_i plays a fundamental part in the discussion of the systems. From its definition it can be stated that

$$\mu_i \equiv G(i) \tag{1.17}$$

For a thermodynamic system at pressure p and temperature T the chemical potentials $\mu_i(p, T)$ of its components are written with the help of activity functions a_i as

$$\mu_i(p, T) = \mu_i^{\text{ref}}(p, T) + RT \ln a_i \tag{1.18}$$

The individual potential $\mu_i(p, T)$ is explained as the result of the interplay of energies, that of the component i as a pure compound (or in infinitely dilute solution or in another convenient reference state) $\mu_i^{\text{ref}}(p, T)$, and that caused by the presence of component i at a given concentration in the mixture; and by interactions of the component i with the other components present. The term a_i is called the activity of component i:

$$a_i = f_i N_i \tag{1.18a}$$

In (1.18a), N_i is the mole fraction of the component i and f_i is the activity coefficient that takes into account the energies of interaction.

The *reference potential* $\mu_i^{\text{ref}}(p,T)$ may refer either to the pure phase of the component i or its infinitely dilute solution.[†] As the chemical potential $\mu_i(p,T)$ itself is an invariant energy content, this leads to the postulation of two kinds of activity coefficients; one of them, f_i, used with reference to the pure phase of the component i,[‡]

$$\mu_i(p,T) = \mu_i^*(p,T) + RT\ln N_i + RT\ln f_i \tag{1.19a}$$

and the other, f_{0i}, used with reference to the infinitely dilute solution of component i[‡]

$$\mu_i(p,T) = \mu_i^\infty(p,T) + RT\ln N_i + RT\ln f_{0i} \tag{1.19b}$$

Because the activity coefficients f_i and f_{0i} are a measure of the interaction of component i in the mixture, they equal unity for the state of reference concerned:

$$\lim_{N_i \to 1} f_i = 1 \qquad \lim_{N_L \to 1} f_{0i} = 1 \tag{1.20a,b}$$

where N_L is the mole fraction of the solvent.

The activity coefficients can either be measured (e.g., vapor pressure experiments or related procedures) or calculated from theoretical models (e.g., Debye–Hückel theory). Combination of (1.16) and (1.17) with (1.19a) or (1.19b), respectively, yields the expressions

$$\bar{H}_i = H_i^* - RT^2 \left(\frac{\partial \ln f_i}{\partial T}\right)_{p,n_i} \qquad \bar{H}_i = H_i^\infty - RT^2 \left(\frac{\partial \ln f_{0i}}{\partial T}\right)_{p,n_i} \tag{1.21a,b}$$

$$\bar{S}_i = S_i^* - R\ln N_i f_i - RT\left(\frac{\partial \ln f_i}{\partial T}\right)_{p,n_j}$$

$$\bar{S}_i = S_i^\infty - R\ln N_i f_{0i} - RT\left(\frac{\partial \ln f_{0i}}{\partial T}\right)_{p,n_j} \tag{1.22a,b}$$

$$\bar{G}_i = G_i^* + RT\ln N_i f_i \qquad \bar{G}_i = G_i^\infty + RT\ln f_{0i} \tag{1.23a,b}$$

[†]Other reference states, such as the ideal 1 m solution, are not considered in this introductory section. For detailed information, see Refs. 41, 52 and 53.

[‡]In accordance with the recommendations of the IUPAC Commission on Symbols, Terminology, and Units[54] the following superscripts are used: * pure substance, ∞ infinite dilution, and ° or ⊖ standard.

which can be used to derive calculable expressions for ΔH_R, ΔS_R, and ΔG_R by inserting in (1.11), (1.13), and (1.14).

1.2.4 REFERENCE STATE AND STANDARD CONDITIONS[†]

In (1.21) to (1.23) the meaning of the reference potentials is

$$H_i^* = -T^2 \left[\frac{\partial (\mu_i^*/T)}{\partial T} \right]_{p,n_i} \qquad H_i^\infty = -T^2 \left[\frac{\partial (\mu_i^\infty/T)}{\partial T} \right]_{p,n_i} \quad (1.24\text{a,b})$$

$$S_i^* = -\left(\frac{\partial \mu_i^*}{\partial T} \right)_{p,n_i} \qquad S_i^\infty = -\left(\frac{\partial \mu_i^\infty}{\partial T} \right)_{p,n_i} \quad (1.25\text{a,b})$$

$$G_i^* = \mu_i^* \qquad\qquad G_i^\infty = \mu_i^\infty \quad (1.26\text{a,b})$$

This means that the reference potentials depend on the same pressure p and temperature T as the total system. Only if a reaction is conducted at $p = 1$ atm and $T = 298.15$ K can the *reference potentials* be replaced by the tabulated *standard potentials* without approximations.

Standard enthalpies H_i^0 of pure compounds are the enthalpies of reaction $\Delta_f H_{298}^0(i)$ of building up these compounds from their elements under standard conditions.[‡]

By convention, chemical elements themselves in their most stable modification have a zero standard enthalpy of formation.

EXAMPLE. $\Delta_f H_{298}^0(H_2) = 0$; $\Delta_f H_{298}^0(S_{\text{rhomb}}) = 0$ but $\Delta_f H_{298}^0(S_{\text{mono}}) = 71$ cal mol^{-1}.

The inclusion of ions in aqueous solution in the thermodynamic system of standard enthalpies is based on the enthalpies of their infinitely dilute aqueous solution. The different state of reference will be denoted by the superscript $\ominus$ in order to simplify the following discussion: $H_i^\ominus = \Delta_f H_{298}^\ominus(i)$. In the extended system of standard enthalpies of formation, $\Delta_f H_{298}^\ominus[H^+(\text{aq})] = 0$.

EXAMPLE. This system leads to the following way of defining the standard enthalpy of formation of an ion Cl$^-$(aq): One mole HCl(g) is dissolved under standard conditions in sufficient water to yield an infinitely dilute solution. The reaction

$$\text{HCl(g)} + \text{H}_2\text{O(l)} \rightarrow \text{H}^+(\text{aq}) + \text{Cl}^-(\text{aq})$$

takes place with a heat effect that is the heat of formation of the pair of ions

---

[†]For the superscripts see the footnote on p. 9.

[‡]Δ_f is the symbol for "formation"; the superscript ^{0} indicates $p = 1$ atm; and the subscript $_{298}$ indicates $T = 298.15$ K.

H$^+$(aq) + Cl$^-$(aq) and is the sum of the heat of solution $\Delta_s H_{298}^0$ of HCl and the heat of formation of HCl, which is completely dissociated in this process:

$$\Delta_f H_{298}^{\ominus}[\text{H}^+(\text{aq}) + \text{Cl}^-(\text{aq})] = \Delta_s H_{298}^0 + \Delta_f H_{298}^0(\text{HCl})$$

$$= -17960 \text{ cal mol}^{-1} - 22060 \text{ cal mol}^{-1} = -40020 \text{ cal mol}^{-1}$$

Since $\Delta_f H_{298}^{\ominus}[\text{H}^+(\text{aq})] = 0$ by convention, the entire enthalpy is attributed to the ion Cl$^-$(aq): $\Delta_f H_{298}^{\ominus}[\text{Cl}^-(\text{aq})] = -40.02$ kcal mol^{-1}.

The temperature dependence of the heat of formation for pure compounds $\Delta_f H_T^0(i)$ and for components in aqueous solutions $\Delta_f H_T^{\ominus}(i)$ is derived from our knowledge of the molar heat capacities $C_p^0(i)$ and $C_p^{\ominus}(i)$,

$$\Delta_f H_T^0(i) = \Delta_f H_{298}^0(i) + \int_{298}^T C_p^0(i) \, dT$$

$$\Delta_f H_T^{\ominus}(i) = \Delta_f H_{298}^{\ominus}(i) + \int_{298}^T C_p^{\ominus}(i) \, dT$$

(1.27a, b)

The temperature-dependent functions $C_p^0(i)$ and $C_p^{\ominus}(i)$ of compounds and aqueous solutions have been tabulated.

The pressure dependence of the heat of formation can be neglected within the pressure range of our discussion. Thus we can use the relationships

$$H_i^*(p, T) = \Delta_f H_T^0(i) \quad \text{and} \quad H_i^{\infty}(p, T) = \Delta_f H_T^{\ominus}(i)$$

as sufficient approximations. Further, the expressions $(\partial \ln f_i / \partial T)_{p, n_i}$ and $(\partial \ln f_{0i} / \partial T)_{p, n_i}$ must be obtained from other methods, unless their values are negligibly small within the range of the employed approximations. Taking into account these remarks, we can calculate ΔH_R for a given reaction with the help of (1.11) and tabulated values.

Standard Entropies and Standard Gibbs Energies. Tables of standard entropies $S_{298}^0(i)$ and $S_{298}^{\ominus}(i)$ for the same conditions ($p = 1$ atm, $T = 298.15$ K) as in standard enthalpies of formation are available. Dependence on temperature of the molar entropy follows from

$$S_T^0(i) = S_{298}^0(i) + \int_{298}^T C_p^0(i) \frac{dT}{T} \qquad S_T^{\ominus}(i) = S_{298}^{\ominus}(i) + \int_{298}^T C_p^{\ominus}(i) \frac{dT}{T}$$

(1.28a, b)

The standard Gibbs energy of a compound i can be calculated either according to (1.15) using $\Delta_f H_{298}^0(i)$ or $\Delta_f H_{298}^{\ominus}(i)$, and $S_{298}^0(i)$ or $S_{298}^{\ominus}(i)$; or from tabulated $\Delta_f G_{298}^0(i)$ values [$\Delta_f G_{298}^{\ominus}(i)$ values]. In the latter system different standard conditions must be used.

The temperature dependence of the Gibbs energy is also known via (1.15), in which $\Delta_f H_T^0(i)[\Delta_f H_T^\ominus(i)]$ and $S_T^0(i)[S_T^\ominus(i)]$ must be inserted for this purpose; the pressure dependence is given by integrating (1.16b), which leads to

$$G_i^*(p) = G_i^0 + V_i^*(p-1) \qquad G_i^\infty(p) = G_i^\ominus + V_i^\infty(p-1)$$

when the molar volume V_i^* or V_i^∞ is handled independently of pressure. This is possible for condensed phases (liquids, solutions, and solids); further, for these $V_i^*(p-1) \ll G_i^0$ and $V_i^\infty(p-1) \ll G_i^\ominus$ for moderate pressures. For gaseous compounds the relationship

$$\mu_i(p,T) = \mu_i^0(T) + RT \ln p_i \tag{1.29}$$

can be obtained from (1.19a) and (1.16b); p_i is the partial pressure (strictly, the fugacity) of the gaseous compound i.

In accordance with general usage we shall henceforth omit the symbols used in thermodynamic functions for differentiating the systems of reference, when the context makes it clear which particular case is meant. The chemical potential, for example, is written

$$\mu_i = \mu_i^0 + RT \ln a_i \tag{1.30}$$

where μ_i^0 is used both for the pure phase as the state of reference, $\mu_i^0 = \Delta_f G_T^0(i)$, and for the infinitely dilute solution, $\mu_i^0 = \Delta_f G_T^\ominus(i)$; a_i is the corresponding activity $a_i = N_i f_i$ or $a_i = N_i f_{0i}$ (it is replaced by p_i for gaseous compounds).

1.2.5 CHEMICAL EQUILIBRIUM

The condition for chemical equilibrium is $\Delta G_R = 0$ in (1.14). Introducing the chemical potential according to (1.30), we get

$$[z_1 \mu^0_{P_1} + \cdots + z_n \mu^0_{P_n}] - [x_1 \mu^0_{A_1} + \cdots + x_n \mu^0_{A_n}]$$

$$= -RT \ln \frac{a_{P_1}^{z_1} a_{P_2}^{z_2} \cdots a_{P_n}^{z_n}}{a_{A_1}^{x_1} a_{A_2}^{x_2} \cdots a_{A_n}^{x_n}} \tag{1.31}$$

The left-hand side of (1.31) is the Gibbs energy of reaction of the process according to (1.1) at a temperature T and $p = 1$ atm; it can be calculated from tabulated data according to the rules given above. This means that the equilibrium constant

$$K_a^{(N)} = \frac{a_{P_1}^{z_1} a_{P_2}^{z_2} \cdots a_{P_n}^{z_n}}{a_{A_1}^{x_1} a_{A_2}^{x_2} \cdots a_{A_n}^{x_n}} \tag{1.32}$$

of the process is calculable in the same way. The superscript $^{(N)}$ indicates the use of mole fractions for measuring concentrations; similarly, we have $K_a^{(c)}$ (1.3) with concentrations measured in moles per liter and the activity coefficients y_i corresponding to this measure of concentration.

1.2.6 PRACTICAL APPLICATION OF THERMODYNAMIC PRINCIPLES IN THERMOMETRIC TITRATION

Since it is possible to calculate in advance the heat of reaction and the equilibrium state for a titration reaction, we can frequently decide whether a particular titration reaction is suitable for the thermometric method and we can establish the best concentration range for the reactants. Information is also available about the possibility of titrating more than one component of a mixture in a single titration. In planning a titration it is generally sufficient to calculate first approximations of ΔH_R and ΔG_R^0. This means that standard data can be used for titrations at room temperatures.

EXAMPLE. Examination of the usefulness of the reaction of neutralization $H^+(aq) + OH^-(aq) \rightleftharpoons H_2O(l)$ as a titration reaction. Standard data show that the equilibrium lies almost completely on the side of the reaction product. We can, therefore, expect a sharp end point. From the heat of reaction of -13 kcal mol^{-1}, a temperature rise of about $\Delta T = 0.70°C$ is calculated for a sample of 50 ml of 0.001 N HCl. The titration can consequently be conducted with the help of very simple instrumentation. The result of the titration (see Chapter 4) follows: The end point is indicated with an accuracy of 0.2%, based on the HCl concentration used. The reaction enthalpy derived from the titration diagrams is -13.35 kcal mol^{-1}. In a similar way it can be predicted from standard data that a titration of K_2HPO_4 (aq) with KOH(aq) will not give a sharp end point in spite of adequate heat of reaction ($\Delta H_R = -6.5$ kcal mol^{-1}) on account of the equilibrium of the reaction (third dissociation constant of phosphoric acid $K_3 = 1.2 \cdot 10^{-12}$); it can be predicted also that a titration of sulfuric acid with caustic soda solution will not give a sharp break in the titration curve corresponding to NaHSO$_4$.

The inclusion of infinitely dilute nonaqueous ionic solutions in the system of standard enthalpies is theoretically possible, but data are lacking so far. An important result of the thermometric method is that it can provide these enthalpies under standard conditions and also, as we shall see later, the entropy of reaction and the Gibbs energy of reaction. Recent applications of the method, for example, in investigating important reactions in biology, are therefore highly interesting (Chapter 8).

A further application of the method is to verify reaction mechanisms by comparing calculated and experimental data. For example, the reaction of phenol with dimethylformamide is described in the existing literature as yielding a 1:1 molecular compound, whereas thermometric titration shows it to be a two-step equilibrium.[55]

Thermometric titration is a suitable method for the investigation of complexes and molecular compounds (Chapter 8). This is one of its recent fields of application in direct relationship with thermodynamics. It is the only method that can provide the enthalpies of reaction and formation constants from the same experiment. However, it must be remembered that (1.11) does not yield standard values; in precise determinations of $\Delta H°$ of ion complexes, for example, an extrapolation to zero ionic strength is necessary (cf. Ref. 41).

A related method is the use of thermometric titration for determining heats of mixing (Chapter 3).

Another investigation directly related to thermodynamics is the application of thermometric titration to distribution systems in order to determine the number of ligands of extracted compounds.[56]

1.3 ELECTROCHEMICAL PRINCIPLES

Many titration reactions take place in aqueous or nonaqueous electrolyte solutions. The following sections point out their qualities when the solutions are of importance for the thermometric method.

1.3.1 IONOPHORIC AND IONOGENIC ELECTROLYTES

The former classification of electrolytes into strong and weak electrolytes on the basis of their conductivity in aqueous solutions leads to difficulties when nonaqueous solutions are taken into account. More fruitful is a classification, on a structural basis, into true and potential electrolytes[57] or ionophores and ionogenes.[58]

Substances that, in their pure phases, contain ions in ionic lattice crystals, such as the alkali halides, are true electrolytes (ionophores). The group of potential electrolytes (ionogenes), such as acetic acid, produces ions only in the course of a chemical reaction with the molecules of the solvent.

True electrolytes are completely dissociated in solution; in solvents of sufficiently low dielectric constant they associate into pairs and even more complex aggregates[59,60,61] as follows:

$$Na^+ + Cl^- \underset{CH_3COCH_3}{\rightleftharpoons} [Na^+Cl^-]^0 \tag{1.33}$$

The ionogenic reaction of potential electrolytes with the solvent is incomplete:

$$CH_3COOH + H_2O \rightleftharpoons CH_3COO^- + H_3O^+ \quad (1.34)$$

In nonaqueous solvents the formation of associates of potential electrolytes in addition to undissociated compounds is also observed,[62,63]

$$HClO_4 + CH_3COOH \rightleftharpoons [CH_3COOH_2^+ \cdot ClO_4^-]^0 \rightleftharpoons CH_3COOH_2^+ + ClO_4^- \quad (1.35)$$

Water also is a potential electrolyte, yielding its ions in a reaction of autoprotolysis

$$H_2O + H_2O \rightleftharpoons OH^- + H_3O^+ \quad (1.36)$$

Both types of reaction, association and dissociation, are defined with the help of thermodynamic equilibrium constants K_A (association constant) or K_D (dissociation constant):

$$K_A = \frac{[Na^+Cl^-]}{[Na^+][Cl^-]} \frac{y_0}{y_+ y_-} \quad (1.37)$$

$$K_D = \frac{[CH_3COO^-][H^+]}{[CH_3COO^-]} \frac{y_+ y_-}{y_0} \quad (1.38)$$

In (1.37) and (1.38), y_0 is the activity coefficient of the undissociated or associated compound; y_+, that of the cation, and y_-, that of the anion. Constants for autoprotolysis in reactions of the type of (1.36) are defined as

$$K_s = [OH^-][H^+] \quad (1.39)$$

According to general usage, the concentration of the solvent (in the present case $[H_2O]$), is not written.

It is impossible to distinguish associates, such as $[Na^+Cl^-]^0$, from undissociated molecules by means of thermodynamic methods.

1.3.2 ACTIVITY COEFFICIENT

The state of reference of an electrolyte in solution is always its infinitely dilute solution. If we consider the simplest case of a true electrolyte Y of composition $K_{\nu_+}^{z_+} A_{\nu_-}^{z_-}$ (K^{z+} signifies a cation of z_+ valency and A^{z-} an anion of z_- valency), which is completely dissociated in a given solvent,

$$K_{\nu_+}^{z_+} A_{\nu_-}^{z_-} \rightarrow \nu_+ K^{z+} + \nu_- A^{z-} \quad (1.40)$$

the electrical neutrality of the electrolyte leads to the equation

$$\nu_+ z_+ = |\nu_- z_-| \tag{1.41}$$

With an infinitely dilute solution as standard, the chemical potentials for cation and anion are given by

$$\mu_+ = \mu_+^0 + RT \ln c_+ y_+ \tag{1.42}$$

$$\mu_- = \mu_-^0 + RT \ln c_- y_- \tag{1.43}$$

Only the chemical potential of the electrolyte Y can be measured, however.

$$\mu_Y = \mu_Y^0 + RT \ln c_Y y_Y \tag{1.44}$$

Electrochemistry usually accounts for this fact with the help of the mean concentration $c_\pm$ and the mean activity coefficient $y_\pm$ of the electrolyte Y:

$$c_\pm = [\nu_+^{\nu_+} \nu_-^{\nu_-}]^{1/\nu} c_Y \qquad \nu = \nu_+ + \nu_- \tag{1.45}$$

$$y_\pm = [y_+^{\nu_+} y_-^{\nu_-}]^{1/\nu} \tag{1.46}$$

As a thermodynamic property, the chemical potential is additive so that we have

$$\mu_Y = \nu_+ \mu_+ + \nu_- \mu_- \tag{1.47}$$

The combination of (1.47), (1.45), and (1.46) gives the following equation:

$$\mu_Y = \mu_Y^0 + \nu RT \ln c_\pm y_\pm \tag{1.48}$$

This equation enables one to express the chemical potential of the completely dissociated electrolyte Y to be expressed through the activities of its ions. This is of particular interest because ion activity can be deduced from a theoretical discussion of models (e.g., Debye–Hückel theory).

If the electrolyte Y is incompletely dissociated, that is, it exists in a state of free ions and associates or undissociated molecules, again only the chemical potential μ_Y can be measured. It is composed of the three parts μ_+, μ_-, and μ_{assoc} ($\mu_{undissoc}$).

If Y, after complete dissociation in the solvent, yields ion aggregates of the type $A = K_{w_+}^{z_+} A_{w_-}^{z_-}$, we get[59]

$$y_\pm^\nu = y_\pm'^\nu \alpha^{\nu_+} \left[1 - (1-\alpha) \frac{w_-}{w_+} \frac{\nu_+}{\nu_-} \right]^{\nu_-} \tag{1.49}$$

in which $y_\pm'$ is the mean activity coefficient in a hypothetical complete dissociation and α the degree of dissociation. In the simplest case, in which the aggregate A is identical with the compound Y itself, we obtain from (1.49) putting $w_+ = \nu_+$ and $w_- = \nu_-$,

$$y_\pm^\nu = y_\pm''^\nu(\alpha^{\nu_+}\alpha^{\nu_-}) \qquad (1.50)$$

and for the electrolytes of (1.33) or (1.34) with $\nu_+ = \nu_- = 1$ and $\nu = 2$,

$$y_\pm = \alpha y'_\pm \qquad (1.51)$$

The mean activity coefficient $y_\pm$ of Y is a function of both the interaction expressed by $y'_\pm$ and the degree of dissociation α. The calculation of $y'_\pm$, that is, of the activity coefficient resulting from the interaction of free ions, is based on the Debye–Hückel theory. At very low electrolyte concentration, the limiting law†

$$\log y_\pm = -\mathrm{A}|z_+ z_-|\sqrt{I} \qquad (1.52)$$

is valid. At higher concentration, the activity coefficient is expressed by the following formula:

$$\log y_\pm = -\frac{\mathrm{A}|z_+ z_-|\sqrt{I}}{1 + \mathrm{B}a\sqrt{I}} \qquad (1.53)$$

If we introduce the physical constants in (1.52) and (1.53) derived from the theory, we obtain $\mathrm{A} = 1.8246 \cdot 10^6 (\epsilon T)^{-\frac{3}{2}}$ mol$^{-\frac{1}{2}}$ liter$^{\frac{1}{2}}$ K$^{\frac{3}{2}}$ and $\mathrm{B} = 50.29 \cdot 10^8 (\epsilon T)^{\frac{1}{2}}$ cm^{-1} mol$^{-\frac{1}{2}}$ liter$^{\frac{1}{2}}$ K$^{\frac{3}{2}}$. ϵ is the dielectric constant of the solvent, T the absolute temperature, and I the ionic strength of the solution.

For further information on this question and on other extensions of the limiting law for higher concentration ranges, the reader is referred to the relevant monographs.[59,60]

1.3.3 AQUEOUS ELECTROLYTE SOLUTIONS

If an acid AH [a mol liter^{-1}] and a base BOH [b mol liter^{-1}] are dissolved in water, the following equilibrium is established:

$$\mathrm{AH} + \mathrm{BOH} \underset{\mathrm{H_2O}}{\rightleftarrows} \mathrm{A}^- + \mathrm{H}^+ + \mathrm{B}^+ + \mathrm{OH}^- \qquad (1.54)$$

†Equations (1.52) and (1.53) are valid only for completely dissociated electrolytes or for the dissociated part of an electrolyte. For the partially associated (or undissociated) electrolyte $\log y_\pm = -\mathrm{A}|z_+ z_-|\sqrt{I} + \log \alpha$ according to (1.51).

The dissociation constant of the acid is

$$K_d^{(AH)} \approx \frac{[A^-][H^+]}{[AH]} = k_a \qquad (1.55)$$

The dissociation constant of the base is

$$K_d^{(BOH)} \approx \frac{[B^+][OH^-]}{[BOH]} = k_b \qquad (1.56)$$

The constant of autoprotolysis of water is

$$K_s = [H^+][OH^-] \qquad (1.57)$$

Also

$$a = [A^-] + [AH] \qquad (1.58)$$

$$b = [B^+] + [BOH] \qquad (1.59)$$

and we must take into account the electrical neutrality of the solution

$$[H^+] + [B^+] = [OH^-] + [A^-] \qquad (1.60)$$

From (1.54) to (1.60) we get

$$[H^+] + \frac{b}{1 + K_s/k_b[H^+]} = \frac{K_s}{[H^+]} + \frac{a}{1 + [H^+]/k_a} \qquad (1.61)$$

If a bivalent acid $A^{(2)}H_2$ with the dissociation constants

$$\frac{[AH^-][H^+]}{[AH_2]} = k_{a_1} \quad \text{and} \quad \frac{[A^{2-}][H^+]}{[AH^-]} = k_{a_2}$$

replaces AH under otherwise unchanged conditions, we obtain instead of (1.61)

$$[H^+] + \frac{b}{1 + K_s/k_b[H^+]} = \frac{K_s}{[H^+]} + \frac{a(2 + [H^+]/k_{a_2})}{1 + [H^+]/k_{a_2} + [H^+]^2/k_{a_1}k_{a_2}} \qquad (1.62)$$

In the most general case of the aqueous solution of several multivalent acids $A^{(v_i)}H_{v_i}$ and bases $B^{(w_i)}(OH)_{w_i}$, the following equation is obtained:

$$[H^+] + \sum_{i=1}^{n} b_i \frac{w_i + \sum_{j=1}^{w_i}(w_i - j)Y_{ij}}{1 + \sum_{j=1}^{w_i} Y_{ij}} = \frac{K_s}{[H^+]} + \sum_{i=1}^{m} a_i \frac{v_i + \sum_{j=1}^{v_i}(v_i - j)X_{ij}}{1 + \sum_{j=1}^{v_i} X_{ij}}$$

$$(1.63)$$

with

$$X_{ij} = \left[\frac{1}{k_{av_i}} \frac{1}{k_{a(v_i-1)}} \frac{1}{k_{a(v_i-2)}} \cdots \frac{1}{k_{a(v_i-j+1)}} \right] [H^+]^j \quad (1.63a)$$

$$Y_{ij} = \left[\frac{1}{k_{bw_i}} \frac{1}{k_{b(w_i-1)}} \frac{1}{k_{b(w_i-2)}} \cdots \frac{1}{k_{b(w_i-j+1)}} \right] \frac{K_s^j}{[H^+]^j} \quad (1.63b)$$

Equation (1.63) is transformed into (1.62) if

$$X_1 = \frac{1}{k_{a_2}}[H^+] \quad X_2 = \frac{1}{k_{a_2}} \frac{1}{k_{a_1}}[H^+]^2$$

and

$$Y_1 = \frac{1}{k_b} \frac{K_s}{[H^+]}$$

Equation (1.61) enables $[H^+]$ to be defined for every degree of titration b/a; the corresponding degree of dissociation α of the acid AH is likewise yielded:

$$\alpha = \frac{[A^-]}{[AH]+[A^-]} = \frac{k_a}{k_a+[H^+]} \quad (1.64)$$

if calculated by using (1.55) and its defining equation.

In a similar way we derive $[H^+]$ in the titration of a bivalent acid according to (1.62), and the degree of dissociation for the two protons as

$$\alpha = \frac{[AH^-]}{[AH_2]+[AH^-]+[A^{2-}]} = \frac{k_{a_1}[H^+]}{[H^+]^2+k_{a_1}[H^+]+k_{a_1}k_{a_2}} \quad (1.65a)$$

$$\beta = \frac{[A^{2-}]}{[AH_2]+[AH^-]+[A^{2-}]} = \frac{k_{a_1}k_{a_2}}{[H^+]^2+k_{a_1}[H^+]+k_{a_1}k_{a_2}} \quad (1.65b)$$

and so forth.

A similar system of equations is used by Holmes and Williams[64] in their "mathematical approach to stepwise enthalpy titrations by use of a computer."

1.4 TITRATIONS IN NONAQUEOUS SOLVENTS

The discussion in Section 1.3.3 is based on the Arrhenius acid-base theory, which defines an acid as a compound that releases protons in aqueous solution, whereas a base releases hydroxyl ions. The theory can be extended to account for amphiprotic solvents similar to water, the molecules of which suffer autoprotolysis according to (1.36), forming lyonium ions (cations of the reaction of autoprotolysis) and lyate ions (anions),

$$NH_3 + NH_3 \rightleftarrows NH_4^+ + NH_2^- \qquad (1.66a)$$

$$ROH + ROH \rightleftarrows ROH_2^+ + RO^- \qquad (1.66b)$$

$$RCOOH + RCOOH \rightleftarrows RCOOH_2^+ + RCOO^- \qquad (1.66c)$$

Neutralization in NH_3 as a solvent in the titration of NH_4Cl with $NaNH_2$ is analogous to acid-base titration in water.

If we write SH for the molecule of an amphiprotic solvent, (1.66a), (1.66b), and (1.66c) become

$$SH + SH \rightleftarrows SH_2^+ + S^- \qquad (1.67)$$

The autoprotolysis constant is

$$K_s = [SH_2^+][S^-] \qquad (1.68)$$

Neutralization can almost always be interpreted through Brönsted's acid-base concept.[65,66] According to Brönsted, acids are proton donors and bases are proton acceptors. Neutralization consists simply in an exchange of protons that changes the given acid S_1 and base B_1 into the weaker acid S_2 and the weaker base B_2,

$$S_1 + B_1 \rightleftarrows S_2 + B_2 \qquad (1.69)$$

for example,

$$H_3O^+ + NH_3 \rightleftarrows NH_4^+ + H_2O \qquad (1.70)$$

The reaction proceeding according to (1.69) is accompanied by a loss of free energy. Acids may be cations (H_3O^+), molecules (CH_3COOH), or anions (HSO_4^-), and the same is true of bases. One and the same ion can be both acid and base (i.e., amphoteric), depending on the kind of reaction,

$$H_3SO_4^+ \rightarrow H_2SO_4 \rightarrow HSO_4^- \rightarrow SO_4^{2-}$$

where H_2SO_4 and HSO_4^- are amphoteric; the other two compounds are pure acid and pure base, respectively.

There are other reactions defined as neutralizations even though no protons are exchanged, for example, the reaction of $AlCl_3$ with dioxane.

The more comprehensive theory of Lewis[67] explains these cases by defining an acid as the acceptor of a free electron pair and a base as the donor. The Brönsted theory can be included in this more comprehensive theory when the proton functions as the acceptor.

Two effects relevant for titration can be observed in amphiprotic solvents. First, the solution of an acid AH or a base B in a solvent such as SH

$$AH + SH \rightleftharpoons A^- + SH_2^+ \qquad (1.71)$$

$$B + SH \rightleftharpoons BH^+ + S^- \qquad (1.72)$$

can lead to a shifting of the equilibrium toward lyonium or lyate ions, which are the strongest possible acids and bases in the given solvent. The original differences in strengths of the various acids and bases in the mixture thereby disappear. A leveling of this sort takes place, for instance, with HCl and H_2SO_4 in water, where protolysis gives H_3O^+ only. If the differences in the strengths of acids are to be demonstrated, solvents that are less basic than water, such as glacial acetic acid, must be used.

The second effect is due to a reaction of the solvent with the salt resulting from neutralization

$$A^- + SH \rightleftharpoons AH + S^- \qquad (1.73)$$

$$BH + SH \rightleftharpoons SH_2^+ + B \qquad (1.74)$$

If these equilibria do not lie distinctly on the left-hand side of the equation, the resulting salts are split by protolysis. This can render an end-point indication of the equivalence point impossible. An example would be the titration of boric acid in aqueous solution by means of a ΔG titration.

Aprotic solvents show no peculiar features in this respect. They are generally good solvents for neutralization reactions.

REFERENCES

1. P. M. Dean and O. O. Watts, *J. Am. Chem. Soc.*, **46**, 855 (1924).
2. P. Dutoit and E. Grobet, *J. Chim. Phys.*, **19**, 324 (1922).
3. G. W. Ewing, *Instrumental Methods of Chemical Analysis*, McGraw-Hill, New York, 1954.
4. J. Jordan and T. G. Alleman, *Anal. Chem.*, **29**, 9 (1957).
5. H. H. Willard, L. L. Merritt, Jr. and J. A. Dean, *Instrumental Methods of Analysis*, 3rd. ed., Van Nostrand, Princeton, 1958, p. 594.
6. J. Jordan, in I. M. Kolthoff and P. J. Elving, *Treatise on Analytical Chemistry*, Part I, Vol. 8, Interscience, New York, 1968, p. 5175.
7. I. Poulsen and J. Bjerrum, *Acta Chem. Scand.*, **9**, 1407 (1955).

8. J. Jordan, *Rec. Chem. Prog.*, **19**, 193 (1958).
9. J. M. Bell and C. F. Cowell, *J. Am. Chem. Soc.*, **35**, 49 (1913).
10. H. D. Richmond and J. E. Merreywether, *Analyst*, **42**, 273 (1917).
11. T. Somija, *J. Soc. Chem. Ind.*, **35**, 135 (1932).
12. T. Somija, *J. Soc. Chem. Ind.*, **33**, 174 (1930). (Suppl.)
13. V. R. Gray and P. F. Whelan, *Chem. Ind.*, 126 (1955).
14. R. E. Press, *Tappi*, **48**, 464 (1965).
15. J. Jordan, P. T. Pei and R. A. Javick, *Anal. Chem.*, **35** 1534 (1963).
16. N. D. Weiner and A. Felmeister, *Anal. Chem.*, **38**, 515 (1966).
17. A. B. De Leo and M. J. Stern, *J. Pharm. Sci.*, **53**, 993 (1964).
18. A. B. De Leo and M. J. Stern, *J. Pharm. Sci.*, **54**, 911 (1965).
19. A. B. De Leo and M. J. Stern, *J. Pharm. Sci.*, **55**, 173 (1966).
20. A. E. Beezer and A. K. Slawinski, *Talanta*, **18**, 837 (1971).
21. J. R. Chipperfield, L. Rossi-Bernardi, and F. J. W. Roughton, *J. Biol. Chem.*, **242**, 777 (1967).
22. E. B. Smith and P. W. Carr, *Anal. Chem.*, **45**, 1688 (1973).
23. J. Jordan, *Chimia*, **17**, 101 (1963).
24. W. F. Hillebrand, G. E. F. Lundell, H. A. Bright, and J. I. Hoffman, *Applied Inorganic Analysis*, 2nd ed., Wiley, New York, 1953, p. 753.
25. J. Jordan and E. J. Billingham, Jr., *Anal. Chem.*, **33**, 120 (1961).
26. J. Jordan and P. W. Carr, in R. S. Porter and J. F. Johnson, *Analytical Calorimetry*, Plenum, New York, 1968, p. 203.
27. P. W. Carr, in L. Meites, *Critical Reviews in Analytical Chemistry*, Chemical Rubber Co., Cleveland, Ohio, Vol. 2, 1972, p. 491.
28. F. J. W. Roughton, in A. Weissberger, *Technique of Organic Chemistry*, Interscience, New York, Vol. 8, Part 2, 1963, p. 758.
29. R. H. Müller, *Ind. Eng. Chem.*, Anal. ed., **13**, 667 (1941).
30. G. W. Ewing, *Instrumental Methods of Chemical Analysis*, McGraw-Hill, New York, 1954.
31. S. T. Zenchelsky, *Anal. Chem.*, **32**, 289 R (1960).
32. M. Harmelin, *Chim. Anal.*, **44**, 153 (1962).
33. J. Jordan, *J. Chem. Educ.*, **40**, A5 (1963).
34. J. Jordan and G. J. Ewing, in L. Meites, *Handbook of Analytical Chemistry*, McGraw-Hill, New York, 1963, Sect. 8, p. 3.
35. L. S. Bark, *Ind. Chem.*, **1963**, 545.
36. I. Sajó, *Kém. Közl.*, (Budapest), **26**, 119 (1966).
37. C. B. Murphy, *Encycl. Ind. Chem. Anal.*, **3**, 672 (1966).
38. A. J. Streiff, *Ann. N.Y. Acad. Sci.*, **137**, 375 (1966).
39. F. Štráfelda and J. Kroftová, *Sb. Vys. Skoly Chem.-Tech. Praze*, Anal. Chem., **2**, 167 (1967).
40. H. J. V. Tyrrell, *New Sci.*, **37**, 300 (1968).
41. J. J. Christensen and R. M. Izatt, in H. A. O. Hill and P. Day, *Physical Methods in Advanced Inorganic Chemistry*, Interscience, New York, 1968, p. 538.
42. F. Becker, *Chem. Ing. Tech.*, **41**, 1105, 1060 (1969).
43. I. W. Breman, *Chem. Tech. (Amst.)*, **24**, 161, 201 (1969).

44. I. A. Crisan, *Rev. Chim. (Buchar.)*, **21**, (2), 99 (1970).
45. L. D. Hansen, R. M. Izatt, and J. J. Christensen, in J. Jordan, *New Developments in Titrimetry*, Dekker, New York 1974, p. 1.
46. H. J. V. Tyrrell and A. E. Beezer, *Thermometric Titrimetry*, Chapman and Hall, London, 1968.
47. L. S. Bark and S. M. Bark, *Thermometric Titrimetry*, Pergamon, Oxford, 1969.
48. G. A. Vaughan, *Thermometric and Enthalpimetric Titrimetry*, Van Nostrand Reinhold, New York, 1973.
49. D. W. Rogers and R. J. Sasiela, *Talanta*, **20**, 232 (1973).
50. P. G. Zambonin and J. Jordan, *Anal. Chem.*, **41**, 437 (1969).
51. S. T. Zenchelsky, J. Periale, and J. C. Cobb, *Anal. Chem.*, **28**, 67 (1956).
52. K. Denbigh, F. R. S., *The Principles of Chemical Equilibrium*, 3rd ed., Cambridge University Press, New York, 1971.
53. J. G. Kirkwood and I. Oppenheim, *Chemical Thermodynamics*, McGraw-Hill, New York, 1961.
54. M. L. McGlashan, *Manual of Symbols and Terminology for Physicochemical Quantities and Units*, Butterworths, London, 1970.
55. F. Becker, J. Barthel, N. G. Schmahl, G. Lange, and H. M. Lüschow, *Z. Phys. Chem.* (New Series), **37**, 33 (1963).
56. A. Kettrup and H. Specker, *Z. Anal. Chem.*, **230**, 241 (1967).
57. G. Kortüm, *Lehrbuch der Elektrochemie*, 4th ed., Verlag Chemie, Weinheim, 1966.
58. R. M. Fuoss and F. Accascina, *Electrolytic Conductance*, Interscience, New York, 1959.
59. R. A. Robinson and R. H. Stokes, *Electrolyte Solutions*, Butterworths, London, 1970.
60. H. S. Harned and B. B. Owen, *The Physical Chemistry of Electrolytic Solutions*, Reinhold, New York, 1954.
61. J. Barthel, *Ionen in nichtwässrigen Lösungen*, Dr. Dietrich Steinkopff Verlag, Darmstadt (im Druck).
62. I. M. Kolthoff and S. Bruckenstein, *J. Am. Chem. Soc.*, **78**, 1 (1956).
63. S. Bruckenstein and I. M. Kolthoff, *J. Am. Chem. Soc.*, **78**, 10, 2974 (1956).
64. F. Holmes and D. R. Williams, *J. Chem. Soc.*, Sect. A, **1967**, 729.
65. J. N. Brönsted, *Rec. Trav. Chim. Pays-Bas*, **42**, 718 (1923).
66. J. N. Brönsted, *Chem. Rev.*, **5**, 231 (1928).
67. G. N. Lewis, *Valence and the Structure of Atoms and Molecules*, Chemical Catalog Co., New York, 1923.

CHAPTER

2

THEORETICAL BASIS FOR AN ANALYSIS OF THERMOGRAMS

2.1 THERMOGRAMS

The plot of the result of a thermometric titration as a T-v diagram or T-t diagram is frequently called a thermogram;[1] other names are enthalpogram[2,3] or simply titration diagram. The term T is the temperature observed or registered that has been produced by the addition of a volume v or at the time t of the titration.

In a titration of the type described in the early literature, the reactant B is fed through a burette into the sample of the titrand A, usually contained in a Dewar flask, and the temperature is observed on a highly sensitive thermometer after each of a succession of added equal volumes of the titrant B. This gives a thermogram of the type in Fig. 2.1, which is reproduced from the first investigation of thermometric titration.[4]

In modern thermometric titration, which is predominantly automatic, titrant B is usually added with the help of a thermostated, motor-driven syringe. The titrand A is contained in a calorimeter vessel. The temperature is preferably measured with the help of thermistors. The thermogram is recorded as a T-t diagram. Figure 2.2 reproduces such a thermogram; it is taken from the first experiment using automatic titration apparatus.[5]

The diagram starts with a preperiod AB before the addition of reactant. The titrant is added at point B. The reaction $OH^- + H^+$ stops at point C (end point of the titration). The titrant is added beyond C and section CD is mainly the heat of dilution of 0.1147 N hydrochloric acid fed into the 0.005 NaCl solution.

The form of the thermogram depends on whether the titrand and titrant have the same or a different temperature and whether the titration reaction is endothermic or exothermic (Fig. 2.3).[2,3]

Other forms of thermogram are discussed later in connection with particular thermometric methods. The forms of thermogram illustrated so far contain the essential basic features necessary for presenting a theory of thermometric titration.

A complete thermogram contains the following information:

1. Indication of the end point of the titration; in a multistep process this may include the indication of single steps.

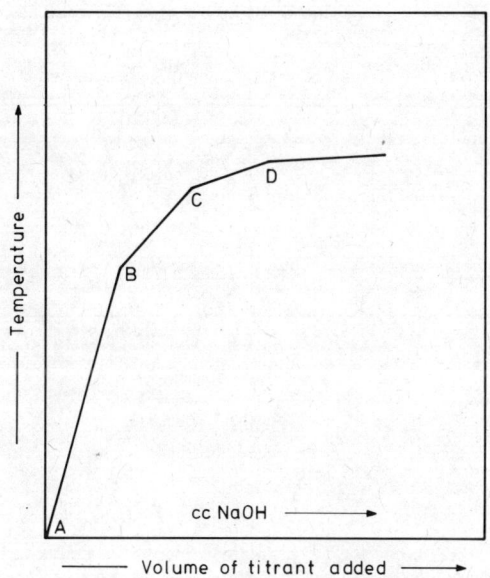

Fig. 2.1. T-v thermogram of a manual thermometric titration, $H_3PO_4 + NaOH$, according to Dutoit and Grobet.[4] (A) H_3PO_4 (beginning of titration); (B) NaH_2PO_4; (C) Na_2HPO_4; and (D) Na_3PO_4.

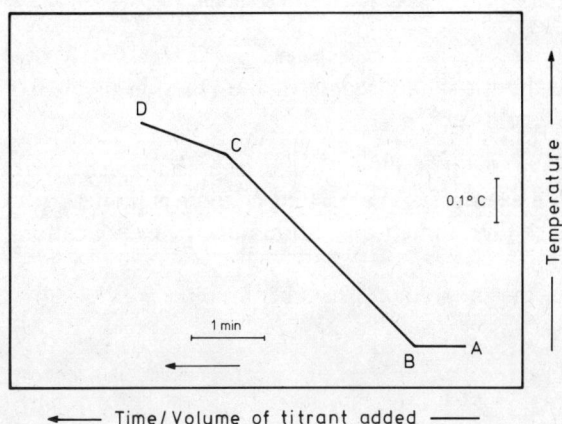

Fig. 2.2. T-t diagram of an automatic thermometric titration: $NaOH + HCl$ according to Linde, Rogers, and Hume.[5] NaOH, 0.005 N; HCl, 0.1147 N.

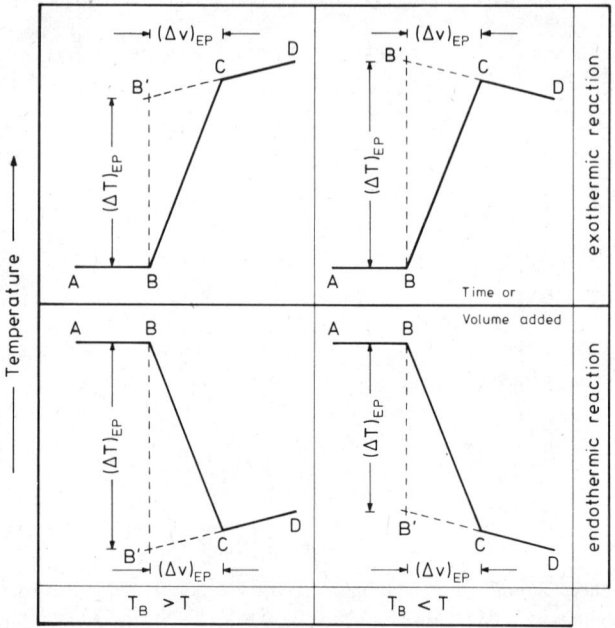

Fig. 2.3. Thermograms under various conditions (using Fig. 17 from Ref. 2). AB, preperiod; BC, reaction period; CD, excess reagent line = dilution period; B, start of the titration; and C, end point of the titration. T_B, temperature of the titrant.

2. Complete thermodynamic data of the reaction or of its various steps: ΔH^0, ΔG^0, and ΔS^0.

This general statement is illustrated in the subsequent chapters with the help of suitable examples. The present chapter is limited to a discussion of the principles of thermometric titration and their application in single-step titration reactions.

A complete thermogram consists of three parts:

1. Preperiod.
2. Titration period.
3. Afterperiod.

In Fig. 2.1 there is no pre- or afterperiod; in Fig. 2.2 there is no afterperiod. It could be obtained if, in Fig. 2.2, the temperature rise were observed for a period about equal to the preperiod after stopping the addition of HCl. If the aim is only to obtain an end-point indication, it is unnecessary to observe the pre- and afterperiods (see Fig. 2.1). When exact thermodynamic data of the titration reaction are sought, however, the complete diagram is necessary.

There are two basic types of thermogram. The first type is that of a titration based on a reaction in which every addition of the titrant is completely and immediately consumed up to the end point. These are the complete titration reactions. All the titrations of Fig. 2.1 to 2.3 are of this type. The complete thermogram has the form of Fig. 2.4. The end point at t_α divides the titration period into a reaction period (dilution of the titrant added and chemical reaction of the diluted titrant) and a dilution period (further dilution of the titrant).

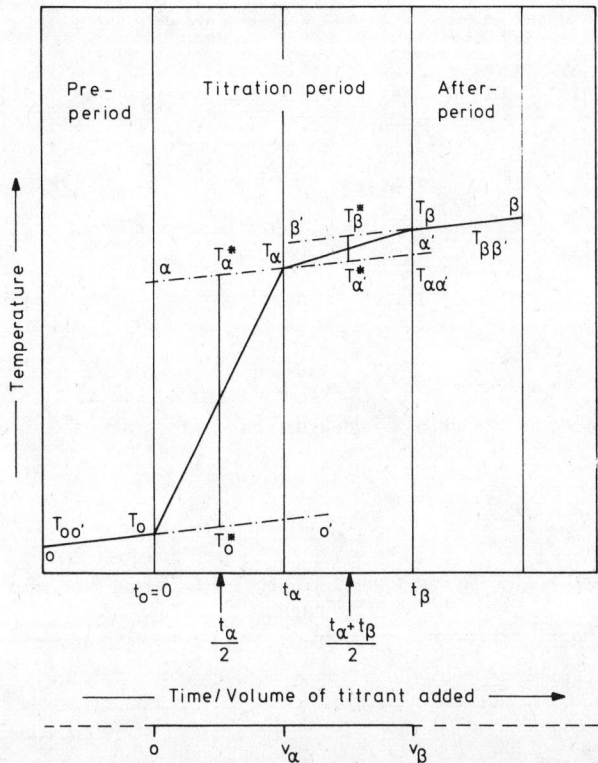

Fig. 2.4. Complete thermogram of a titration with *complete* reaction. End point: $v_\alpha = v'_0 t_\alpha$.

The second type is observed in a titration reaction in which the titrant, irrespective of the quantity added, is not used up completely; A, B, and the end products of the reaction are in a state of equilibrium. The titration reaction is incomplete, and no end point is seen. The thermogram has the form given in Fig. 2.5; there is no division of the titration period into reaction period and dilution period. For the determination of the

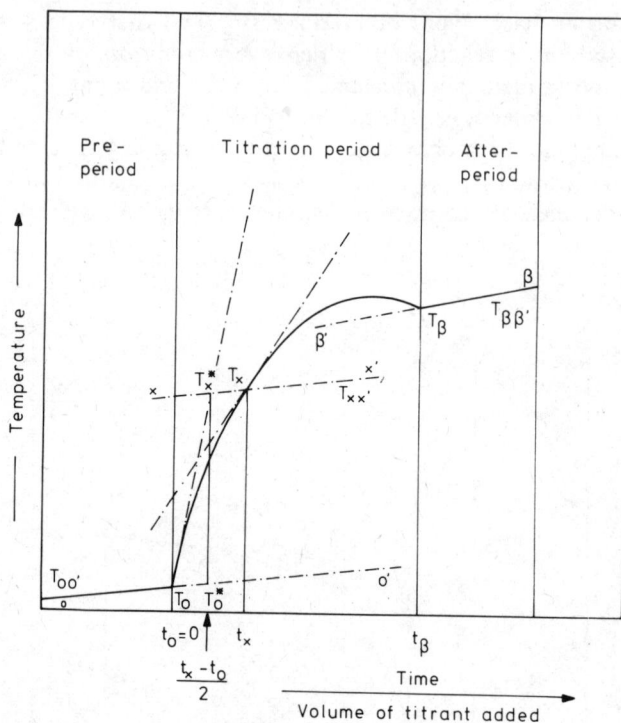

Fig. 2.5. Complete thermogram of a titration with *incomplete* reaction.

TABLE 2.1. Synonymous Terms for the Different Parts of a Thermogram

	Vaughan[6]	Jordan[2]	Carr[8]	Barthel (This work)
1.	Base period	—	Fore period	Preperiod
2.	—	—	Reaction period	Titration period
	Reaction period	Titration branch	—	Reaction period
	Excess titrant period	Excess reagent branch	—	Period of dilution
3.	Post titrant period	—	Post period	Afterperiod

28

thermometric data it is necessary, as a rule, to measure the heat of dilution by adding B to the pure solvent of A in a parallel experiment.

The various expressions used by different authors for parts of a thermogram are collected in Table 2.1.

For an adequate explanation of multistep titration reactions, it may be necessary to combine the two basic types. In the neutralization of a weak acid, there is a superimposition of the neutralization reaction, which is complete, and dissociation of the weak acid, which is an equilibrium reaction.

2.2 THE BASIC EQUATION OF THERMOMETRIC TITRATION

Thermometric titration is a calorimetric method. The thermograms are explained on the basis of the calorimetric equation, which gives the balance of thermal powers of the process. The thermal power W_i, measured at the time $t = t_x$, is related to the heat of reaction Q_i according to

$$W_i(t_x) = \left(\frac{dQ_i}{dt}\right)_{t=t_x} \tag{2.1}$$

To obtain an equation valid for evaluation of a thermogram, each and every effect connected with the implementation of the experiment and influencing the heat content of the calorimeter must be appropriately taken into account.[9,10] This means that the heat balance is no longer restricted to the heat of reaction as in Section 1.2; this becomes the partial power w_R of the total balance (see Section 2.2.5). For this partial effect, the statements of Section 1.2 remain valid.

The special part played by the titrant among the reactants in the titration reaction according to (1.1) is indicated by its characterizing symbol B. The titrant is added stepwise in the course of the titration, whereas the other reactants A_i are in the titration vessel from the very beginning. The titrant is, of course, one of the initial reactants in thermodynamic consideration.

2.2.1 THE OVERALL THERMAL POWER

If a change of temperature dT is measured in the calorimeter vessel of heat capacity C, the overall thermal power W_{tot} during the period of observation dt is given by

$$W_{\text{tot}} = C\frac{dT}{dt} \tag{2.2}$$

The heat capacity C (cal K^{-1}) of the calorimeter is the sum of the following elements: (1) the content of the calorimeter and (2) the calorimeter vessel itself and its component parts, such as the temperature sensor, stirrer, and heater, insofar as these parts undergo temperature change (see Section 1.2.1).

Since the content of the calorimeter changes during titration as a result of addition of the titrant B, the heat capacity in (2.2) is a function of the added quantity of B[†]:

$$C = C_0 + (c_B + \delta)v \qquad (2.3)$$

where C_0 is the heat capacity of the calorimeter with its content of solution A before B is added. The quantity c_B with the dimension (cal K^{-1} liter^{-1}) is a rather unconventional "specific heat," but it has proved to be useful in thermometric titration and is used in the following discussion. The amount of solution B added up to the time of measurement is given as v (liters) in (2.3). The function δ accounts for the fact that, during the process of adding B, further parts of the calorimeter and its components that were not covered by the solution at the time $t = 0$ become immersed and heated. By appropriately constructing the titration calorimeter (see Chapter 9), δ can be made negligible in practice or can be eliminated by suitable methods of evaluation.

Neglecting δ, we obtain

$$C(t) = C(t_0) + c_B \int_{t_0}^{t} \frac{dv}{dt} dt \qquad (2.4)$$

where dv/dt is the velocity (liters sec^{-1}) with which solution B is added.

The values of C_0, c_B, and if necessary, δ, must be known if thermodynamic data are to be derived from a thermogram. Their determination is discussed in Section 3.1.2.

2.2.2 THERMAL POWER OF THE MIXING PROCESS

The addition of the titrant B from the reservoir (burette and syringe), in which it is held at a constant temperature T_B ready for titration of the content of the calorimeter, the temperature T of which changes as the titration proceeds, implies a thermal power during the process of mixing

$$W_M = -c_B(T - T_B)\frac{dv}{dt} \qquad (2.5)$$

[†]In (2.3) the heat capacities of titrand and addition are simply added. This approximation adequately accounts for the reality of the process and leads to simple equations in the following theoretical discussions. The error to be expected from this simplification in determining ΔH_R is usually less than 0.1%.

The thermal power W_M of the physical process of mixing must not be confused with the power of the process of dilution of B in the titrand A, to be discussed later. A thermal power W_M would arise on mixing two samples of pure water of differing temperatures, but there would be no power of dilution.

2.2.3 THERMAL POWER OF THE REACTION

There are always two connected thermal powers. First, there is the power of the reaction itself w_R, defined by the titration reaction; and, second, the power, w_v, of the dilution process resulting from the dilution of solution B:

$$W = w_R + w_v \tag{2.6}$$

To develop an expression for the thermal power of reaction, $w_R(t)$, the chemical reaction equations valid at the point t on the titration curve must be known. Let us take n_A, n_B, and n_P as the mole numbers of the initial reactants A (titrand) and B (titrant) and the reaction product P at a given time. We then get

$$w_R = \frac{dQ_R}{dt} = -\frac{d}{dt}(n_P \Delta H_R) \tag{2.7}$$

if ΔH_R (cal mol^{-1}) is the molar enthalpy of reaction of the titration reaction, which in the present case is taken to be a single-step reaction

$$x_A A + x_B B \rightleftarrows z_P P \tag{2.8}$$

according to (1.1).

For the limited temperature range usual in thermometric titration, ΔH_R is frequently taken to be independent of temperature:

$$w_R = -\Delta H_R \frac{dn_P}{dt} \tag{2.9}$$

In complex reactions in which reaction products P_i are formed in several steps, (2.9) must be replaced by

$$w_R(t) = \sum_i w_R^{(i)} = -\sum_i \Delta H_R^{(i)} \frac{dn_P^{(i)}}{dt} \tag{2.10}$$

We begin the discussion of a complete reaction on the basis of (2.8). During the titration period, that is, until the equivalence point is reached, every addition of B is used up completely.

$$\frac{dn_P}{dn_B} = \frac{z_P}{x_B} = \nu \qquad (2.11)$$

If 1 mol B produces 1 mol P during the reaction, $\nu = 1$. When the number of moles n_B is expressed by the concentration b (mol liter^{-1}) and the volume added from the burette is v, we obtain from (2.9)

$$w_R = -\Delta H_R b \frac{dv}{dt} \qquad (2.12)$$

Using the molar enthalpy of dilution H_v, which characterizes the heat developed during the dilution of the solution B from concentration b to concentration zero, we get during the titration period

$$w_v = -H_v b \frac{dv}{dt} \qquad (2.13)$$

It must be remembered that the functions ΔH_R and H_v refer to a given solvent and to the ionic strength and temperature of the sample at time t. In thermometric titrimetry it is frequently enough for practical purposes to use the approximation (see Section 1.2.4): $\Delta H_R \approx \Delta H_R^0$.

In complete titration reactions $\Delta H_R = 0$ naturally during the period of dilution. Equation (2.13) can be used as an approximation for the period of dilution, even though, in this section of the diagram, the dilution of solution B does not reach the concentration zero of B. A more exact formula would be

$$w_v = -[H_v - H_v']b \frac{dv}{dt} \qquad (2.14)$$

instead of (2.13).[11] Here H_v is the same function as before and H_v' is the heat of dilution for the process

B (concentration in the calorimeter)→B (concentration zero)

Therefore H_v' is a function of the volume v added or, in the T-t diagram, a function of time.

In the case of titrations with incomplete reaction (see Fig. 2.5) equation (2.6) in the nondeveloped form is frequently used in the following discussion.

2.2.4 THERMAL POWERS CAUSED BY THE CALORIMETER

The thermal power of the calorimeter W_K comprises w_N, due to the temperature gradient between the calorimeter content and environment, and w_0, fed into the calorimeter by the stirrer:

$$W_K = w_N + w_0 \qquad (2.15)$$

The power w_N is defined through Newton's law of cooling

$$w_N = -C\kappa(T-\theta) \qquad (2.16)$$

As in (2.2), C is the time-dependent heat capacity of the calorimeter, κ (sec^{-1}) the cooling constant, and θ the surrounding temperature. When the titration vessel is situated in a thermostat, the constant temperature of this thermostat is taken as θ.

The power of the stirrer w_0 (cal sec^{-1}) can be included as a constant with a suitably constructed calorimeter:

$$w_0 = \text{const} \qquad (2.17)$$

Methods for the determination of κ and w_0 are given in Section 3.1.2. For a detailed discussion of the thermal powers caused by the calorimeter see Chapter 9.

2.2.5 THE BASIC EQUATION OF THERMOMETRIC TITRATION

The thermal power balance of the process of titration at time t is

$$W_{\text{tot}} = W_M + W + W_K$$

By including the relevant equations from Sections 2.2.1 to 2.2.4, the generally valid basic equation of thermometric titration is derived as a differential equation

$$C\frac{dT}{dt} = w_R(t) + w_v(t) + w_0 - c_B(T-T_B)\frac{dv}{dt} - C\kappa(T-\theta) \qquad (2.18)$$

The exact solution of this linear homogeneous differential equation with the initial conditions $t = t_0$ and $T = T_0$ is

$$C(T-T_0) = e^{-\kappa t}\int_{t_0}^{t}(w_R + w_v)e^{\kappa t}dt + c_B(T_B - \theta)e^{-\kappa t}\int_{t_0}^{t}\frac{dv}{dt}e^{\kappa t}dt$$

$$- c_B(T_0 - \theta)\int_{t_0}^{t}\frac{dv}{dt}dt + [1 - e^{-\kappa(t-t_0)}]\left[\frac{w_0}{\kappa} - C(t_0)(T_0 - \theta)\right] \qquad (2.19)$$

It contains no approximations. Equation (2.19) is the integral form of the basic equation of thermometric titration. It describes the heat balance for an interval between two arbitrarily chosen points of time t_0 and t of the titration

$$Q_{tot} = C(T - T_0)$$

Depending on the aim and the method used, (2.19) can be applied occasionally in a simplified form; such simplifications sometimes imply an increasing degree of approximation, however.

A few simplifications that can be realized through suitable construction of the calorimeter and do not impair the accuracy of measurements should be mentioned here. It is easy to keep the temperature of the reservoir (burette or syringe) of solution B constant and equal to the temperature θ of the thermostat of the calorimeter. It suffices to surround the reservoir with a jacket connected with the circulation of the calorimeter thermostat: $T_B = \theta$. Further, the titrant B can be added with a constant velocity $dv/dt = v'_0$. This can be accomplished either by ejecting the solution B from a motor-driven burette or syringe at a constant speed, or by adding step by step equal volumes from the reservoir at equal time intervals. In the former case, the result is a really constant velocity v'_0, which is represented by a linear function between added volume v and time t; in the second case, the result is a "step function" between v and t (see Fig. 3.3). Without going into a detailed discussion of the technical control of addition of v, it can be stated that either way of adding the titrant justifies the assumption that $dv/dt = v'_0$.

Under these conditions, which can be realized in the experimental setup, it follows from (2.19) that

$$C(t)(T - T_0) = e^{-\kappa t}\int_{t_0}^{t}(w_R + w_v)e^{\kappa t}dt - c_B(T_0 - \theta)v'_0(t - t_0)$$
$$+ [1 - e^{-\kappa(t - t_0)}]\left[\frac{w_0}{\kappa} - C(t_0)(T_0 - \theta)\right] \qquad (2.20)$$

The application of (2.20) is difficult because of the integral expression it contains. A practicable solution can be found via the transformation with the help of the mean value theorem of integral calculus

$$\int_{t_0}^{t}(w_R + w_v)e^{\kappa t}dt = e^{\kappa\zeta(t - t_0)}\int_{t_0}^{t}(w_R + w_v)dt \qquad 0 \leq \zeta \leq 1$$

Under the usual conditions for thermometric titrations ($\kappa \sim 10^{-4} \text{sec}^{-1}$, $t < 1000$ sec), ζ can be set as $\frac{1}{2}$ and the expansion of the exponential function can be truncated by neglecting the term $\kappa^2 t^2/12$. We then obtain from (2.20)

$$[C(T-\theta) - C_0(T_0-\theta)][1 + \tfrac{1}{2}\kappa(t-t_0)]$$

$$= \int_{t_0}^{t}(w_R + w_v)dt + [w_0 - C_0\kappa(T_0-\theta)](t-t_0) \quad (2.21)$$

Equation (2.21) can be handled without difficulty.

The remaining integral function expresses the integral heat effects of the reaction $Q_R(t)$ and of the dilution $Q_v(t)$,

$$Q_R(t) + Q_v(t) = \int_{t_0}^{t}(w_R + w_v)dt \quad (2.22)$$

$$Q_R(t) = \int_{t_0}^{t} w_R \, dt \quad (2.22a)$$

and

$$Q_v(t) = \int_{t_0}^{t} w_v \, dt \quad (2.22b)$$

The derivation of the basic equation of thermometric titration presented in this section is that proposed by Barthel, Becker, and Schmahl.[9,10] On this basis Christensen, Izatt, Hansen, and Partridge[11] proposed and integrated the differential equation (2.18) with w_v defined as in (2.14) and with the time-dependent heat capacity in Newton's law of cooling.

$$C(T - T_0) = \int_{t_0}^{t}[w_0 - C\kappa(T-\theta)]dt$$

$$+ \int_{t_0}^{t} w_R \, dt - c_B \int_{t_0}^{t}(T - T_B)\frac{dv}{dt}dt + \int_{t_0}^{t} w_v \, dt \quad (2.23)$$

Papoff and Zambonin[12] used (2.12) and (2.13) to obtain the same differential equation (2.18). In a later paper[13] they started from another equation for the heat balance and again reached (2.18), with the restrictions of a constant power of dilution and of a constant heat capacity. The influence of δ according to (2.3) was taken into account. The resulting integral function was presented as the exact solution of the differential equation.

2.3 EVALUATION OF TITRATION DIAGRAMS

2.3.1 MATHEMATICAL DESCRIPTION OF THERMOGRAMS OF TITRATIONS WITH INCOMPLETE TITRATION REACTION

The mathematical description of the most general case of a titration diagram is of the type of Fig. 2.5, which is obtained with an incomplete titration reaction. The description may use a system of differential equations or the corresponding integral functions.

The differential equations for the pre- and afterperiods follow from (2.18) when only the heats of stirring and of cooling are considered for the heat balance, and when $C = C_0$ for the preperiod and $C = C_\beta = C_0 + (c_B + \delta)v_\beta$ for the afterperiod, since there is no addition of titrant.

Expressed as a differential equation, the thermogram of a titration with incomplete reaction can be written as

1. Preperiod ($t \leq 0$):

$$C_0 \frac{dT}{dt} = w_0 - C_0 \kappa (T - \theta) \qquad (2.24a)$$

2. Titration period ($0 \leq t \leq t_\beta$):

$$C \frac{dT}{dt} = w_R + w_V + w_0 - c_B(T - \theta)\frac{dv}{dt} - C\kappa(T - \theta) \qquad (2.24b)$$

3. Afterperiod ($t \geq t_\beta$):

$$C_\beta \frac{dT}{dt} = w_0 - C_\beta \kappa (T - \theta) \qquad (2.24c)$$

The system of equations (2.24) is based on the assumption that the addition of the titrant starts at $t_0 = 0$ and ceases at $t = t_\beta$. It is also assumed that the burette containing the titrant B is controlled by the same thermostat as the titration vessel with the titrand A: $T_B = \theta$.

The integration of (2.24) gives the following:

1. Preperiod ($t \leq 0$):

$$C_0(T - T_0) = w_0 t - C_0 \kappa t (T_0 - \theta) \qquad (2.25a)$$

2. Titration period ($0 \leq t \leq t_\beta$):

$$[C(T - \theta) - C_0(T_0 - \theta)](1 + \tfrac{1}{2}\kappa t) = \int_0^t (w_R + w_v) dt + w_0 t - C_0 \kappa t (T_0 - \theta) \qquad (2.25b)$$

3. Afterperiod ($t \geq t_\beta$):

$$C_\beta (T - T_\beta) = w_0(t - t_\beta) - C_\beta \kappa (t - t_\beta)(T_\beta - \theta) \qquad (2.25c)$$

In (2.25a and c) it is assumed that there is only a small temperature change during the pre- and after periods. This allows a linear approximation of the exponential functions of (2.24a and c). Equation (2.25b) is identical with (2.21) and is, therefore, based on the same assumptions.

2.3.2 MATHEMATICAL DESCRIPTION OF THERMOGRAMS OF TITRATIONS WITH COMPLETE TITRATION REACTION

The more special case of a thermogram of a titration with complete reaction is discussed in the following section. The example chosen is that of a titration with a single and complete reaction, such as that of HCl and NaOH (see Fig. 2.4).

The equations for pre- and afterperiods remain unchanged. The end point of the titration $t = t_\alpha$ separates the titration period into a reaction period ($0 \leq t \leq t_\alpha$) and a dilution period ($t_\alpha \leq t \leq t_\beta$). If we take into account (2.12) to (2.14), the differential equations for these periods are for the preperiod ($t \leq 0$)

$$C_0 \frac{dT}{dt} = w_0 - C_0 \kappa (T - \theta) \quad (2.26a)$$

for the reaction period ($0 \leq t \leq t_\alpha$)

$$C \frac{dT}{dt} = -(\Delta H_R + H_v) b \frac{dv}{dt} + w_0 - c_B(T - \theta) \frac{dv}{dt} - C\kappa(T - \theta) \quad (2.26b)$$

for the period of dilution ($t_\alpha \leq t \leq t_\beta$)

$$C \frac{dT}{dt} = -(H_v - H'_v) b \frac{dv}{dt} + w_0 - c_B(T - \theta) \frac{dv}{dt} - C\kappa(T - \theta) \quad (2.26c)$$

and for the afterperiod ($t \geq t_\beta$)

$$C_\beta \frac{dT}{dt} = w_0 - C_\beta \kappa (T - \theta) \quad (2.26d)$$

The integration of these equations gives

for the preperiod ($t \leq 0$)

$$C_0(T - T_0) = w_0 t - C_0 \kappa t (T_0 - \theta) \quad (2.27a)$$

for the reaction period ($0 \leq t \leq t_\alpha$)

$$[C(T-\theta) - C_0(T_0 - \theta)][1 + \tfrac{1}{2}\kappa t] = -(\Delta H_R + H_v) b v'_0 t + w_0 t - C_0 \kappa t (T_0 - \theta) \quad (2.27b)$$

for the period of dilution ($t_\alpha \leq t \leq t_\beta$)

$$[C(T-\theta) - C_\alpha(T_\alpha - \theta)][1 + \tfrac{1}{2}\kappa(t - t_\alpha)]$$
$$= -(H_v - H'_{v,m}) b v'_0 (t - t_\alpha) + w_0(t - t_\alpha) - C_\alpha \kappa (t - t_\alpha)(T_\alpha - \theta) \quad (2.27c)$$

and for the afterperiod ($t \geqslant t_\beta$)

$$C_\beta(T - T_\beta) = w_0(t - t_\beta) - C_\beta\kappa(t - t_\beta)(T_\beta - \theta) \tag{2.27d}$$

Equations (2.26) have been integrated on the assumption of constant speed of addition $dv/dt = v_0'$. A mean value $H_{v,m}'$ has been taken for the heat of dilution H_v', which varies during the course of the addition of the titrant over the whole period of dilution. In a titration of 0.05 N HCl with 1 N NaOH, H_v' would increase from zero at $t = t_\alpha$ to about 100 cal mol^{-1} at $t = t_\beta$. This means that $H_{v,m}'$ could even be neglected within a normally desired accuracy.

It is frequently difficult to find a suitable function for evaluating the heat of dilution. The following sections deal with this problem.

2.3.3 CALCULATION OF THE END POINT AND AN APPROXIMATE ENTHALPY OF REACTION

Whereas the pre- and afterperiods, and usually also the period of dilution of a complete titration reaction, give straight lines in the thermogram, the reaction period does not, as can be seen from (2.27b). The deviation results from the cooling of the system, characterized by the factor $[1 + \tfrac{1}{2}\kappa t]$, and from the dependence on time of the heat capacity equations (2.3) and (2.4).

$$C = C_0\left(1 + \frac{c_B v_0'}{C_0} t\right) \tag{2.28}$$

The first factor is more or less negligible in the present context, but the second is of considerable importance with poorly soluble titrants, which means that a large amount of the solution B has to be added in the course of the titration. The curvature resulting from the change in the heat capacity equation (2.28) can be eliminated by a modified plot (see Fig. 2.6) with

$$C_0 \Delta T_{\text{corr}} = C_0(T_{\text{corr}} - \theta) = C(T - \theta) \tag{2.29}$$

Such time-consuming plots are usually not necessary. The exact methods of evaluation do account for the curvature. Plotting of the type of (2.29), however, is resorted to in cases of complex reactions in order to find out whether the curvature of the titration curve is due to the changing heat capacity or to a thermal power of reaction that has not yet been respected, or whether there is an equilibrium reaction.

Three cases of end-point indication of a titration reaction must be considered.[14] The relevant diagrams, which are given as Fig. 2.7, illustrate

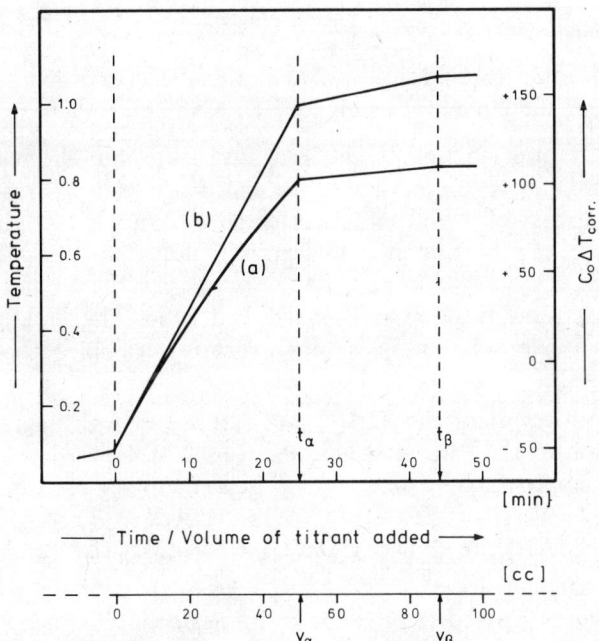

Fig. 2.6. Titration curves: (a) direct; (b) corrected according to (2.29). Titration of $AgNO_3$ ($V_0 = 200$ ml, $a = 0.05\ M$) with NaCl ($b = 0.2\ M$) at 25°C.

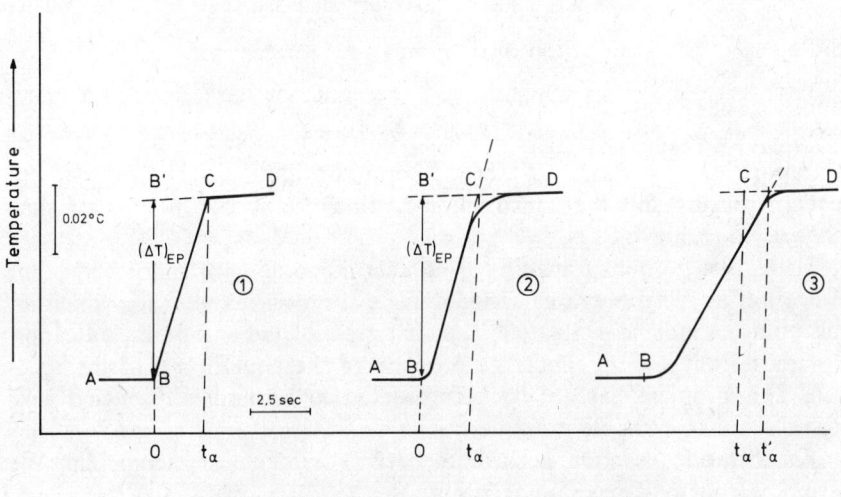

Fig. 2.7. Thermograms under various conditions.[14] (1) Reaction *fast* and *complete*: HCl + NaOH; (2) reaction *incomplete* and *fast*: H_3BO_3 + NaOH; and (3) reaction *complete* and *slow*: R—NO_2 + NaOH.

the following cases:

(1) The titration reaction is *complete* and *fast*. The end point t_α is a sharp break of the curve, for example, HCl+NaOH.

(2) The titration reaction is *fast* and *incomplete*, but the reaction equilibrium is so far on the side of the end products of the reaction that a small excess of titrant effects essentially complete reaction. The end point t_α can be obtained by extrapolation, for example, H_3BO_3+NaOH.

(3) The titration reaction is *complete* but *slow*. The end point found through extrapolation, t'_α, is not correct, for example, RNO_2+NaOH.

The thermograms of Fig. [2.7(1) and (2)] are typical diagrams for the determination of the end point (see also Fig. 2.3). Nevertheless, they can be used for approximately determining the enthalpy of reaction;

$$(\Delta T)_{EP} = \frac{1}{C_0}(-\Delta H_R bv'_0 t_\alpha) = -\frac{b}{C_0}\Delta H_R (\Delta v)_{EP} \qquad (2.30)$$

This equation is derived (2.27 b and c) by neglecting the heats of stirring and cooling and the specific heat c_B of the titrant for the reaction period $(0 < t \leq t_\alpha)$

$$C_0(T - T_0) = -(\Delta H_R + H_v)bv'_0 t \qquad (2.31a)$$

and also $H'_{v,m}$ for the period of dilution $(t_\alpha \leq t \leq t_\beta)$

$$C_0(T - T_\alpha) = -H_v bv'_0(t - t_\alpha) \qquad (2.31b)$$

Extrapolation of the line $\overline{CB'}$ of the period of dilution to the point where reaction begins and the construction of line $\overline{BB'} = \Delta T$ in Fig. 2.7 thus leads us to (2.30).

There is a gradual transition from thermograms with sharp end-point indication at t_α to diagrams of incomplete titration reaction for which an end point cannot be evaluated. The sharpness of the end-point indication is—apart from kinetic effects—a problem of the equilibrium of the reaction. This problem has not been considered so far and is discussed next with the aid of a simple example.

The titration reaction according to (2.8) is taken as incomplete. We assume a course of reaction according to

$$A + B \rightleftharpoons AB$$

encountered in the titration of Lewis acids, for example, in which the initial reactants A and B produce in their incomplete reaction the adduct AB as the final product.

$$K_c = \frac{[AB]}{[A][B]} \qquad (2.32)$$

The titrand A of concentration $a = n_A^0/V_0$ [mol liter^{-1}], is titrated with B (b[mol liter^{-1}]). In the reaction vessel we obtain at time t after the continuous addition of B a total of n_A^0 mol A and n_B^0 mol B, of which

$$n_A = n_A^0 - n_{AB} \qquad (2.33a)$$

$$n_B = n_B^0 - n_{AB} \qquad (2.33b)$$

have not reacted with A and B, respectively. We assume that of the total amount n_B^0 a fraction α reacts to form AB,

$$n_{AB} = \alpha n_B^0 \qquad (2.33c)$$

From (2.32) it follows that

$$\frac{K_c}{V} = \frac{\alpha n_B^0}{n_A^0(1-\alpha\beta)n_B^0(1-\alpha)} \qquad (2.34)$$

with

$$V = V_0 + v_0' t \qquad (2.34a)$$

and

$$\beta = \frac{n_B^0}{n_A^0} = \frac{bv_0' t}{aV_0} \qquad (2.34b)$$

The degree of association α is derived from (2.34), after introducing $\gamma = (K_c n_A^0/V)^{-1}$ by solution of the quadratic equation

$$\alpha^2 \beta - \alpha(\beta + \gamma + 1) + 1 = 0$$

The integral heat effect Q_R at the time t of the titration is, according to (2.9) and (2.33c),

$$Q_R = -n_{AB}\Delta H_R = -(\alpha b v_0' \Delta H_R)t \qquad (2.35)$$

In titration reactions with the complete reaction of B, $\alpha = 1$ in every addition up to the end point of the titration; Q_R is, therefore, a linear function of t with interval $0 \leq t \leq t_\alpha$. Then α becomes zero at the end of the reaction and remains zero for $t > t_\alpha$, resulting in the break at the end point.

In his discussion of titrations with the incomplete reaction of B, Tyrrell[15] combines (2.35) and (2.34b) and introduces the dimensionless variable

$$\xi = -\frac{Q_R}{aV_0\Delta H_R} \tag{2.36}$$

designating a "reduced heat output"; he discusses the function $\xi = \xi(\beta) = \xi(n_B^0/n_A^0)$

$$\xi = \alpha \frac{n_B^0}{n_A^0}$$

as a function of the parameter γ (Fig. 2.8). A curve with $\gamma = 0$ is that of a complete titration reaction. The end point of every reaction would be at $n_B^0 = n_A^0$, that is, at $\beta = 1$. The family of curves in Fig. 2.8 demonstrates the role of the equilibrium constant in giving a sharp end point and shows whether and to what extent an extrapolation of the end point is reasonable in cases of incomplete reaction. It also shows in which cases the enthalpy of reaction can be deduced from the initial slope of the titration curves[15] by employing (2.35). (The exact meaning of the initial slope is discussed in Sections 2.3.4 and 2.3.5.)

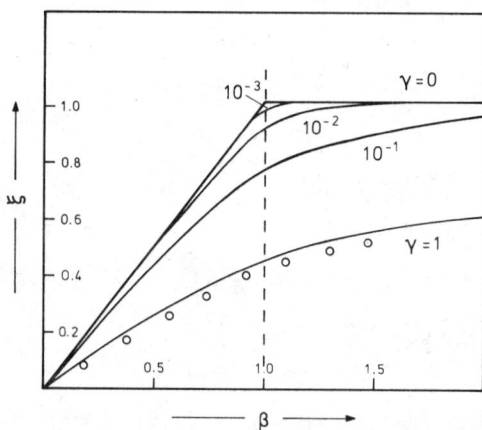

Fig. 2.8. Reduced thermograms for varying values of γ.[15] The upper curve refers to the limiting case of a complete reaction ($\gamma = 0$); the other lines are calculated for the values of γ indicated. The points marked (o) are calculated from the data of Christensen et al.[8] for the reaction $HPO_4^{2-} + OH^- \rightleftharpoons PO_4^{3-} + H_2O$ and correspond to an experiment in which the initial concentration of Na_2HPO_4 was 0.009464 M.[15]

A thermogram of the type of Fig. 2.7(3) is yielded when the rate of the titration reaction, $-d[\mathrm{A}]/dt$, is not large enough compared to the rate of addition of titrant bv_0'. The relation

$$\frac{d[\mathrm{B}]}{dt} = \frac{d[\mathrm{A}]}{dt} + bv_0' \qquad (2.37\mathrm{a})$$

then holds.

For titration reactions that are irreversible and second order,

$$\mathrm{A} + \mathrm{B} \xrightarrow{k} \mathrm{P}$$

and

$$-\frac{d[\mathrm{A}]}{dt} = k[\mathrm{A}][\mathrm{B}] \qquad (2.37\mathrm{b})$$

Combining (2.37a and b) we obtain a differential equation, which, taking into account the initial states of $[\mathrm{A}] = [\mathrm{A}]_0$ and $[\mathrm{B}] = 0$ for $t \leq 0$ (preperiod), has the solution

$$[\mathrm{A}] = [\mathrm{A}]_0 \frac{\exp\left\{kt\left([\mathrm{A}]_0 - \frac{bv_0'}{2}t\right)\right\}}{1 + k[\mathrm{A}]_0 \int \exp\left\{kt\left([\mathrm{A}]_0 - \frac{bv_0'}{2}t\right)\right\} dt} \qquad (2.38)$$

The analysis of data shows that an error of 4% in the end-point indication is still brought about by kinetic interference in the thermogram, even at rate constants k of 60,000 mol^{-1} min^{-1}. Jordan and Carr have developed a method of obtaining useful analytical results in situations where the titration curve is significantly distorted by kinetics.[14]

2.3.4 THE SECTION METHOD

The system of equations (2.27) is basically sufficient for determining the enthalpy of reaction ΔH_R of the titration reaction from a thermogram. In order to simplify the calculations $C(t)$, w_0, and κ could be determined in separate experiments (see Section 3.1.2), and also H_v and H_v'. The coordinates (t_0, T_0), (t_α, T_α), and (t_β, T_β) can be obtained by linear extrapolation from the adjoining parts of the curves. This gives us all the data necessary for solving the problem.

The same goal can be attained with the help of a simple graphical method.[9,10] A linear t_α period, as would result from stopping the addition of the reactant at time t_α (Fig. 2.4) is constructed on the basis of pre- and afterperiods of the thermogram. The slope of the t_α period, $dT_{\alpha\alpha'}/dt$, is defined with those of the pre- and afterperiods, $dT_{00'}/dt$ and $dT_{\beta\beta'}/dt$, using the temperature differences $(T_\alpha - T_0)$ and $(T_\beta - T_0)$ as weight functions:

$$\frac{dT_{\alpha\alpha'}}{dt} = \frac{dT_{00'}}{dt} + \frac{T_\alpha - T_0}{T_\beta - T_0}\left(\frac{dT_{\beta\beta'}}{dt} - \frac{dT_{00'}}{dt}\right) \quad (2.39)$$

The t_α period, a straight line $\alpha\alpha'$ in Fig. 2.4, is both the preperiod of the period of dilution and the afterperiod of the reaction period. It obeys the equation

$$C_\alpha(T - T_\alpha) = -C_\alpha\kappa(t - t_\alpha)(T_\alpha - \theta) + w_0(t - t_\alpha) \quad (2.40)$$

Extrapolation of the two straight lines $00'$ [preperiod of the titration (2.27a)] and $\alpha\alpha'$ [t_α period, (2.40)] for the time $t = t_\alpha/2$ (respective functional values T_0^* and T_α^*) gives

$$C_\alpha(T_\alpha^* - T_\alpha) = C_\alpha\kappa\frac{t_\alpha}{2}(T_\alpha - \theta) - w_0\frac{t_\alpha}{2} \quad (2.41a)$$

$$C_0(T_0^* - T_0) = -C_0\kappa\frac{t_\alpha}{2}(T_0 - \theta) + w_0\frac{t_\alpha}{2} \quad (2.41b)$$

The equation (2.27b) of the reaction period can be rewritten as

$$C(T - T_0) + \left[C\kappa\frac{t}{2}(T - \theta) - w_0\frac{t}{2}\right] + \left[C_0\kappa\frac{t}{2}(T_0 - \theta) - w_0\frac{t}{2}\right]$$
$$= -(\Delta H_R + H_v)bv_0't - c_B(T_0 - \theta)v_0't \quad (2.41c)$$

Taking into account (2.41a) and (2.41b) gives, for $t = t_\alpha$

$$C_\alpha(T_\alpha^* - T_0^*) = -(\Delta H_R + H_v)bv_0't_\alpha - c_Bv_0't_\alpha(T_0^* - \theta) \quad (2.42a)$$

In a similar way, the extrapolation of the straight lines $\alpha\alpha'$ and $\beta\beta'$ [afterperiod (2.27d)] to the point $(t_\beta + t_\alpha)/2$ for the period of dilution (2.27c) gives

$$C_\beta(T_\beta^* - T_{\alpha'}^*) = -(H_v - H_{v,m}')bv_0'(t_\beta - t_\alpha)$$
$$- c_Bv_0'(t_\beta - t_\alpha)(T_{\alpha'}^* - \theta) \quad (2.42b)$$

The quantities T_i^* and $T_{i'}^*$ can be obtained according to Fig. 2.4; $(\Delta H_R + H_v)$ and $(H_v - H'_{v,m})$ are calculated from (2.42a) and (2.42b). In the vast majority of cases $H'_{v,m}$ is negligible so that ΔH_R can be calculated by combining the two equations, taking (2.4) into account:[7]

$$\Delta H_R = \frac{1}{bv'_0}\left[C_\alpha\left(\frac{T_\beta^* - T_{\alpha'}^*}{t_\beta - t_\alpha} - \frac{T_\alpha^* - T_0^*}{t_\alpha}\right) + c_B v'_0(T_\beta^* - T_0^*)\right] \quad (2.43)$$

The advantage of this procedure is the elimination of w_0, κ, and θ, which need not be determined separately.

Equation (2.42a) may be understood as follows: If the entire amount of the titrant $v_\alpha = v'_0 t_\alpha$ of the reaction period $(t_0 \cdots t_\alpha)$ were added at the moment $t_\alpha/2$, an inertia-free temperature sensor would show a jump from T_0^* to T^*_α. Such an imagined process takes place instantaneously and is, therefore, independent of κ, w_0, and θ. It is clear that ΔH_R can only be an approximation because of the value used for the heat of dilution. Whereas the evaluation of the reaction period yields $\Delta H_R + H_v$, that of the dilution period gives $H_v - H'_{v,m}$. The value of $H'_{v,m}$ can either be neglected or introduced as a constant, which can taken from the literature or determined in independent experiments.

The question of whether the enthalpy of reaction ΔH_R is dependent on the ionic strength of the solution must be investigated separately whenever it becomes relevant.

A t_x period can, of course, be constructed for any point t_x of the titration period instead of for the abscissa value t_α. Thus the sectional method can be applied to thermograms of incomplete reactions to give $Q_R(t_x) + Q_v(t_x)$ (see Fig. 2.5). On the basis of the system of equations (2.25) it gives

$$\int_0^{t_x}(w_R + w_v)dt = Q_R(t_x) + Q_v(t_x)$$

$$= C_x(T_x^* - T_0^*) + c_B v'_0 t_x(T_0^* - \theta) \quad (2.44)$$

in which T_0^* and T_x^* are the ordinates of linear extrapolation of the preperiod and of the t_x period at the point $t_x/2$. Since thermograms of incomplete reactions have no period of dilution, the value $Q_R(t_x)$ must be determined with the help of a separate dilution experiment; this consists in filling the reaction vessel with the pure solvent and adding the titrant B at the same speed. By constructing the t_x period in this diagram using the sectional method in analogy to (2.44), the value of $Q_v(t_x)$ is obtained, and, by forming the corresponding differences, finally $Q_R(t_x)$. This evaluation contains approximations similar to those in the titration diagram of complete reactions. In an incomplete reaction, the titrant B added is not consumed completely so that the heat of dilution from the titration diagram and that from the parallel experiment are not exactly the same. This difference may usually be neglected.

The section method as such contains no approximations that are not inherent in the system of equations (2.25).

2.3.5 THE TANGENT METHOD

This method can be applied most fruitfully to thermograms of incomplete titration reactions. If the tangent and t_x period (see Fig. 2.5) are constructed for the point (t_x, T_x), the related expression of $w_R(t_x) + w_v(t_x)$ can be derived easily.

The slope of the t_x period is given by

$$C_x \frac{dT_{xx'}}{dt} = -C_x \kappa (T_x - \theta) + w_0 \qquad (2.45a)$$

and that of the titration curve by (2.24b) as

$$C_x \left(\frac{dT}{dt}\right)_{t_x} = w_R(t_x) + w_v(t_x) + w_0 - c_B(T_x - \theta)v_0' - C_x \kappa(T_x - \theta) \qquad (2.45b)$$

The elimination of w_0 and κ produces

$$C_x \left[\left(\frac{dT}{dt}\right)_{t_x} - \frac{dT_{xx'}}{dt}\right] + c_B v_0'(T_x - \theta) = w_R(t_x) + w_v(t_x) \qquad (2.46)$$

The parallel experiment of adding the titrant to the pure solvent of A gives the analogous function for $w_v(t_x)$ alone. The evaluation of both experiments thus enables $w_R(t_x)$ also to be calculated.

A special part is played at the beginning of the titration period by the initial thermal powers[10] $w_R(t_0)$ and $w_v(t_0)$,

$$C_0 \left[\left(\frac{dT}{dt}\right)_0 - \frac{dT_{00'}}{dt}\right] + c_B v_0'(T_0 - \theta) = w_R(t_0) + w_v(t_0) \qquad (2.47)$$

where $dT_{00'}/dt$ is the slope of the preperiod. The initial slope of the titration period $(dT/dt)_0$ can be determined with sufficient accuracy ($\pm 2\%$) because of the sharp break in the recorded T-t curves and the initially sensible linear gradient. A parallel dilution experiment again provides the value of $w_v(t_0)$, which is proportional to the enthalpy of dilution H_v of the solution B,

$$w_v(t_0) = -bv_0' H_v \qquad (2.48)$$

since at $t = t_0$ the mole fraction N_b in the reaction vessel has the value zero. Thermometric titration can thus provide the heat of dilution directly, whereas the classical methods of calorimetry furnish it only via elaborate extrapolation procedures (see Chapter 4). This is true only if mixing is fast and the establishment of equilibrium equally fast, and if the calorimeter system is sufficiently free of inertia.

In a complete titration reaction the function $w_R(t_0)$ has the value

$$w_R(t_0) = -bv_0' \Delta H_R \tag{2.49}$$

Since w_R is independent of time in such a case, it is advisable to use the more accurate section method for the determination of ΔH_R.

The importance of the tangent method for the determination of $w_R(t_0)$ in incomplete titration reactions is discussed in the next section.

2.3.6 SIMULTANEOUS DETERMINATION OF ΔG_R^0, ΔH_R^0, AND ΔS_R^0 FROM A THERMOGRAM

The possibility of evaluating the complete thermodynamic data of a titration reaction or its parts has been quoted as one of the features of a complete thermogram. The method will be illustrated here with the simple case of a single-step equilibrium reaction.[10] More complex cases, especially multistep reactions, are discussed in the following sections; multistep reactions of complex formation are discussed in Chapter 8.

The problem under discussion is, in a way, the reverse of that discussed in Section 2.3.3: We have the thermogram (Fig. 2.5) of a titration with incomplete reaction and an equilibrium constant according to (2.32). We are looking for a method that furnishes the thermodynamic data K_c (or ΔG^0), ΔH^0, and ΔS^0 of the reaction through evaluation of the thermogram. The equations (2.32), (2.33a), (2.33b), and (2.34a)† define the necessary functions,

$$n_{AB} = n_B^0 - n_B = n_A n_B \frac{K_c}{V} \tag{2.50}$$

$$n_A^0 = n_A + n_{AB} = n_A \left(1 + n_B \frac{K_c}{V}\right) \tag{2.51}$$

and with these

$$\frac{K_c}{V} = \frac{n_B^0 - n_B}{n_B(n_A^0 - n_B^0 + n_B)} \tag{2.52}$$

†The further assumption of Section 2.3.3, introducing the degree of association α with (2.33c), is not retained.

The enthalpy of reaction ΔH_R is the complex forming enthalpy ΔH_{AB}. For the integral heat effect [see (2.35)] we get

$$Q_R(t) = -n_{AB}\Delta H_{AB} = -(n_B^0 - n_B)\Delta H_{AB} \qquad (2.53)$$

and from (2.52)

$$\frac{K_c}{V} = -\frac{Q_R \Delta H_{AB}}{(n_B^0 \Delta H_{AB} + Q_R)(n_A^0 \Delta H_{AB} + Q_R)} \qquad (2.54)$$

In order to determine the unknown quantities ΔH_{AB} and K_c, we need the measured values Q_R, n_B^0, and V for two points of time I and II during the titration period. $Q_R^I = Q_R(t_I)$ and $Q_R^{II} = Q_R(t_{II})$ can be determined by the section method; and n_B^{0I}, n_B^{0II}, V^I, and V^{II} can be calculated directly from the addition.

From (2.54), we then obtain the following equation for the determination of ΔH_{AB}:

$$\Delta H_{AB}^2 n_A^0 \left(n_B^{0II} V^I Q_R^I - n_B^{0I} V^{II} Q_R^{II} \right) + \Delta H_{AB} Q_R^I Q_R^{II}$$

$$\times \left[V^I n_B^{0II} - V^{II} n_B^{0I} + (V^I - V^{II}) n_A^0 \right]$$

$$- Q_R^I Q_R^{II} (V^{II} Q_R^I - V^I Q_R^{II}) = 0 \qquad (2.55)$$

The value of ΔH_{AB} from this equation is inserted in (2.54) to give K_c.

Another method is based on the introduction of the initial thermal power $w_R(t_0)$ from (2.47)[10]. The thermal power is defined according to (2.53) and (2.50) as

$$w_R(t) = \frac{dQ_R}{dt} = -\Delta H_{AB}\frac{dn_{AB}}{dt} = -\Delta H_{AB} K_c \frac{d}{dt}\left(\frac{n_A n_B}{V}\right) \qquad (2.56)$$

and thus as

$$w_R(t_0) = \lim_{t \to 0} w_R(t) = -\Delta H_{AB} a K_c \left(\frac{dn_B}{dt}\right)_{t=0} \qquad (2.57)$$

From

$$\left(\frac{dn_B}{dt}\right)_{t=0} = \frac{b v_0'}{1 + a K_c} \qquad (2.58)$$

follows

$$w_R(t_0) = -b v_0' \Delta H_{AB} \frac{a K_c}{1 + a K_c} \qquad (2.59)$$

EVALUATION OF TITRATION DIAGRAMS 49

The two unknown quantities K_c and ΔH_{AB} are, in this case, obtained from determinations of the initial thermal powers, $w_R^I = w_R^I(t_0)$ and $w_R^{II} = w_R^{II}(t_0)$, in two titrations with two different titrand concentrations, a^I and a^{II}. Both the concentration of B and the velocity v_0' of addition of B are the same in the two experiments.

$$K_c = \frac{w_R^I a^{II} - w_R^{II} a^I}{a^I a^{II}(w_R^{II} - w_R^I)} \quad (2.60)$$

Then ΔH_{AB} is obtained by inserting K_c into (2.59).

In dealing with this problem we have neglected the temperature dependence of the constant of formation K_c. The resulting error is within the limits of experimental accuracy for temperature rises of less than 0.5°C.

It must not be forgotten that K_c and ΔH_{AB} are determined with the help of the data of two points of the titration curve that do not fulfill the requirement of equal ionic strength. It is, therefore, more advisable to use instead of (2.32) an equation including the activity coefficient y_i (molarity scale) of the reactants:

$$\frac{K_a}{V} = \frac{n_{AB}}{n_A n_B} \frac{y_{AB}}{y_A y_B} = \frac{n_{AB}}{n_A n_B} Y \quad (2.61)$$

Instead of K_c the function K_a/Y should, therefore, be included in (2.50) to (2.60). With the help of the determination of K_a and ΔH_{AB}^0, we obtain

$$\Delta G_R^0 = -RT \ln K_a \quad (2.62)$$

and hence

$$\Delta S_R^0 = \frac{1}{T}(\Delta H_R^0 - \Delta G_R^0) \quad (2.63)$$

To sum up, all thermodynamic data of the process can be derived from the thermogram.

The equilibrium constants of the reactions studied must be of an order that allows a meaningful application of the method. This problem is discussed in further detail in the following chapters. Titration diagrams of complete reactions ($K_c \to 0$) do not permit the determination of ΔG_R^0 and ΔS_R^0; they can provide only ΔH_R^0.

2.3.7 OTHER METHODS OF EVALUATION

Becker and Grundmann[16] base their determination of the heat of reaction on the equation

$$Q_R(t) + Q_v(t) = Z(t) + \kappa \int_0^t Z(t) dt \quad (2.64)$$

with

$$Z(t) = C_0(T - T_{00'}) + c_B v_0'(T - \theta) \tag{2.64a}$$

The function $Z(t)$ is obtained by extrapolating the preperiod beyond $t=0$, as in Fig. 2.5, $T - T_{00'}$ is the difference between the T value of the titration curve and the $T_{00'}$ value of the extrapolated preperiod at the same point t. Equation (2.64) is based on the complete basic equation of thermometric titration [(2.18) or (2.19)]. The cooling constant κ must be determined in a separate experiment for this method of evaluation.

In this context the investigations of two groups of scientists must be mentioned, namely, those of Papoff et al.[12,13] and those of Christensen et al.[11] They base their equations on the fundamental system of equations (2.18), with variants that we have already encountered in Section 2.2.6.

Papoff and Zambonin[12] describe a graphical evaluation procedure that is based on the developments discussed in Section 2.3.6 and is a further development of the method of Cioffi and Zenchelsky[17] (see also Section 8.2). It allows the simultaneous determination of K_c and ΔH_{AB} from every point of the curve. A separate determination of the heat of dilution and a construction of the t_x period are necessary, however, for the evaluation. The method can be used over a wide range for the determination of constants of formation.

In a further paper[13] the basic statements of Refs. 9 and 10 are taken up and it was demonstrated that, under favorable conditions and using the values $T(t_x)$ and $(\partial T/\partial t)_{tx}$, K and ΔH can be determined for any point on the curve. It must be noted that the evaluation is based on values obtained at the same ionic strength.

Christensen, Izatt, Hansen, and Partridge[11] developed a numerical method for the simultaneous determination of K and ΔH_R from their equation (2.23), derived from the basic equation. The name *entropy titration* was coined by them for methods that allow the simultaneous determination of equilibrium constant and enthalpy of reaction. The system of equations for the theory of entropy titration is that of Section 2.3.6.

The evaluation of the thermogram is based on (2.23). For an application at point $x(t_x, T_x)$ of the titration period it is written as[11]

$$C(T_x - T_0) = \int_0^{t_x} [w_0 + k(\theta - T)]dt + \int_0^{t_x} w_R \, dt + c_B v_x (\theta - T_0)$$

$$- (H_v - H_v')bv_x \tag{2.65}$$

This means that in the cooling term $k = C\kappa$, $v_x = v_0' t_x$, and that (2.14) is used for the thermal power of dilution.

The functions H_v, H'_v, and c_B and the variable heat capacity C must be determined in separate experiments or can be taken from the existing literature.

The integral expression $\int_0^{t_x}[w_0+k(\theta-T)]dt$ is then calculated for a curve that represents an approximation to the titration curve by a series of linear segments $(t_1-t_0, t_2-t_1,\ldots,t_{x+1}-t_x)$ with abscissa steps $t_{i+1}-t_i$ of about 1 min. Since

$$w_0+k(\theta-T)=C_0\frac{dT_{00'}}{dt}+\frac{T-T_0}{T_\beta-T_0}\left(C_\beta\frac{dT_{\beta\beta'}}{dt}-C_0\frac{dT_{00'}}{dt}\right)=S$$

we get for the section $t_{i+1}-t_i$,

$$\int_{t_i}^{t_{i+1}}S\,dt=\tfrac{1}{2}[S(t_{i+1})-S(t_i)](t_{i+1}-t_i)$$

and with n such sections for the whole integral expression

$$\int_0^{t_x}[w_0+k(\theta-T)]dt=\sum_{i=0}^{x}\frac{1}{2}(S_{i+1}+S_i)(t_{i+1}-t_i) \qquad (2.66)$$

The methods for the determination of ΔH_R are accurate to 0.2 to 1% if they take into account the heat of stirring, the heat exchange with the environment, and the heats of mixing and dilution with sufficient approximation.

Methods are frequently found in the literature, especially for the evaluation of thermograms of complete titration reactions, that do not account for all parts of the heat balance contained in (2.18) or (2.19). The degree of approximation of these methods is often satisfactory when the experimental procedure is adequate. The following paragraph presents two such methods that are frequently used.

The initial slope method of Keily and Hume[18] (see also Ref. 2) can be regarded as a special case of the tangent method based on (2.47). If the preperiod is reduced to a horizontal line and if $T_0\approx\theta$, which is easily arranged experimentally, (2.47) and (2.48) lead to

$$C_0\frac{d(T-T_{00'})}{dt}+bv'_0H_v=w_R(t_0) \qquad (2.67)$$

The value of H_v can now be determined by feeding B into the pure solvent of A so that bv'_0H_v of (2.67) is eliminated. The initial gradient of the thermogram then provides $w_R(t_0)$ and, according to (2.49), ΔH_R in a single-step and complete titration reaction. In an incomplete single-step reaction of the type of (2.32), the determination of ΔH_R according to (2.49) depends on the extent to which the condition $aK_c \gg 1$ is fulfilled; (2.59) is needed for an exact determination. In multistep reactions according to (8.10), the initial gradient yields the constant of complex formation and the heat of formation of the first stage of the equilibrium following (2.59). A separate determination of H_v is unnecessary if the titration is carried out as a differential procedure.[19]

The approximation of Jordan[2] goes further than (2.31a) and (2.31b) and contains a term for the heat of mixing and, implicitly, the heat of dilution. The heats of stirring and cooling, which are not respected in this method, become less important when the time of addition is very short, as in automatic titration. If the reagent addition is very fast (a very short titration period), the heats of stirring and cooling become negligibly small and the basic equation (2.19) leads to

$$C(t)(T - T_0) = Q_R(t) + Q_v(t) - c_B v_0'(t - t_0)(T_0 - \theta) \qquad (2.68)$$

This means that the reaction period up to the end point t_α, after complete reaction of the titrand, is given by

$$C_\alpha(T_\alpha - T_0) = -(\Delta H_R + H_v)bv_0't_\alpha - c_B v_0' t_\alpha(T_0 - \theta) \qquad (2.69a)$$

and, neglecting H_v', the period of dilution t_β is given in

$$C_\beta(T_\beta - T_\alpha) = -H_v bv_0'(T_\beta - T_\alpha) - c_B v_0'(T_\beta - T_\alpha)(T_\alpha - \theta) \qquad (2.69b)$$

In the method of extrapolation in Fig. 2.7 we have

$$(\Delta T)_{EP} = (T_\alpha - T_0) - (T_\beta - T_\alpha) \quad \text{and} \quad (\Delta v)_{EP} = v_0' t_\alpha = v_0'(t_\beta - t_\alpha)$$

From (2.69a) and (2.69b), since $C_\beta = C_\alpha + c_B(\Delta v)_{EP}$, we obtain

$$C_\alpha(T_\alpha - T_0) - C_\alpha(T_\beta - T_\alpha) = -\Delta H_R b(\Delta v)_{EP} - c_B(\Delta v)_{EP}(T_0 - T_\alpha)$$

that is,

$$C_\alpha(\Delta T)_{EP} - c_B(\Delta v)_{EP}(T_\beta - T_0) = -\Delta H_R b(\Delta v)_{EP} \qquad (2.70)$$

Equation (2.70) leads to (2.30) if $c_B(\Delta v)_{EP}(T_\beta - T_\alpha)$ is neglected in comparison with $C_0(\Delta T)_{EP}$.

Recently Carr has critically surveyed methods for the determination of reaction enthalpy based on thermograms.[8,20] The fundamental equations of thermometric titration are rewritten using dimensionless variables, and the extrapolation methods are judged with the help of simulated solutions[20] in which the cooling constant, duration of titration, change of the heat capacity caused by the titrant, and characteristic features of the methods of extrapolation are taken into account. Carr evaluates extrapolations at t_0,

t_α, and $t_\alpha/2$ with the help of the extrapolation of preperiod ($\alpha\alpha'$) and afterperiod ($\beta\beta'$) according to Fig. 2.4. Carr's second paper[8] contains a valuable analysis of the precision and sensitivity of thermometric titrations and the effect of thermal parameters on end-point accuracy; it is recommended reading. In this paper Carr applies the general theory of linear titrations of Rosenthal, Jones, and Megargle[21] to thermometric methods.

2.3.8 ANALYSIS OF BINARY MIXTURES AND MULTISTEP REACTIONS

The preceding sections demonstrate the importance of the reaction enthalpy ΔH_R and the equilibrium constant K_a of the titration reaction for analyzing the thermograms of single-step reactions. In a binary mixture there are the following possible combinations:

Case 1: $\Delta H_1 \approx \Delta H_2$

    (a) $\Delta G_1 \approx \Delta G_2$ (b) $\Delta G_1 \neq \Delta G_2$

Case 2: $\Delta H_1 \neq \Delta H_2$

    (a) $\Delta G_1 \approx \Delta G_2$ (b) $\Delta G_1 \neq \Delta G_2$

Mixtures whose components have about the same heat effect in the titration reaction (Case 1), are not suitable for thermometric analysis aimed at evaluating the content of each component. Mixtures with $\Delta G_1 \approx \Delta G_2$ (Cases 1a and 2a) are likewise unsuitable for a Gibbs energy titration.

The presence of two clearly differing reaction enthalpies (Case 2), as, for instance, in Fig. 1.3, is not alone sufficient to produce two equivalence points in the thermogram; ΔG_1 and ΔG_2 (or the respective equilibrium constants) also must differ sufficiently (Case 2b). This case is discussed in Section 4.3 for the titration of mixtures of acids. It is, however, possible to analyze the mixture when $\Delta G_1 \approx \Delta G_2$ (Case 2a), which yields no end point for component 1, provided that n_{tot} and Q_{tot} can be determined,[22,23]

$$n_{tot} = n_1 + n_2 \tag{2.71}$$

$$Q_{tot} = n_1 \Delta H_1 + n_2 \Delta H_2 \tag{2.72}$$

From this it follows that

$$n_1 = \frac{Q_{tot} - n_{tot}\Delta H_2}{\Delta H_1 - \Delta H_2} \tag{2.73}$$

A similar idea was used by Marik-Korda, Buzasi, and Cserfalvi[24] in the development of a method for the simultaneous determination of the individual components in a multicomponent system by titration with nonselective reagents. If the heat of reaction $\Delta H_i^{(j)}$ of each component (i) with each reagent (j) is known and the total heats of reaction $Q^{(j)}$ are measured in DIE determinations (see Chapter 3), the concentrations c_i of n components ($i=1,\ldots,n$) can be determined with n titrations ($j=1,\ldots,n$) of samples with a volume $V^{(j)}$,

$$c_1\Delta H_1^{(1)} + c_2\Delta H_2^{(1)} + \cdots + c_n\Delta H_n^{(1)} = -\frac{Q^{(1)}}{V^{(1)}}$$

$$c_1\Delta H_1^{(2)} + c_2\Delta H_2^{(2)} + \cdots + c_n\Delta H_n^{(2)} = -\frac{Q^{(2)}}{V^{(2)}} \qquad (2.74)$$

$$\vdots$$

$$c_1\Delta H_1^{(n)} + c_2\Delta H_2^{(n)} + \cdots + c_n\Delta H_n^{(n)} = -\frac{Q^{(n)}}{V^{(n)}}$$

The linear system of equations (2.74) must be solved to determine the unknown concentrations $c_1,\ldots,c_n$. The method is illustrated by the simultaneous determination of sulfide and thiosulfate with iodine solution and bromine water as nonselective titrants.

Closely connected with the problem of binary mixtures is the question of the partial neutralization of multivalent acids or bases and the multistage equilibria in reactions of complex formation. This question is discussed in Sections 4.3 and 8.2.

REFERENCES

1. H. J. V. Tyrrell and A. E. Beezer, *Thermometric Titrimetry*, Chapman and Hall, London, 1968.
2. J. Jordan, in I. M. Kolthoff and P. J. Elving, *Treatise on Analytical Chemistry*, Part I, Vol. 8, Interscience, New York, 1968, p. 5175.
3. J. Jordan, *Chimia*, **17**, 101 (1963).
4. P. Dutoit and J. Grobet, *J. Chim. Phys.*, **19**, 324 (1922).
5. H. W. Linde, L. B. Rogers, and D. N. Hume, *Anal. Chem.*, **25**, 404 (1953).
6. G. A. Vaughan, *Thermometric and Enthalpimetric Titrimetry*, Van Nostrand Reinhold, New York, 1973.
7. J. Barthel, N. G. Schmahl, and K. Lenz, *Z. Anal. Chem.*, **233**, 328 (1968).
8. P. W. Carr, in L. Meites, *Critical Reviews in Analytical Chemistry*, Chemical Rubber Co., Cleveland, Ohio, Vol. 2, 1972 p. 491.
9. J. Barthel, F. Becker, and N. G. Schmahl, *Z. Phys. Chem.* (New Series), **29**, 58 (1961).
10. F. Becker, J. Barthel, N. G. Schmahl, G. Lange, and H. M. Lüschow, *Z. Phys. Chem.* (New Series), **37**, 33 (1963).
11. J. J. Christensen, R. M. Izatt, L. D. Hansen, and J. A. Partridge, *J. Phys. Chem.*, **70**, 2003 (1966).

12. P. Papoff and P. G. Zambonin, *Ric. Sci.*, **35**, 93 (1965).
13. P. Papoff, G. Torsi, and P. G. Zambonin, *Gazzetta*, **95**, 1031 (1965).
14. J. Jordan and P. W. Carr, in R. S. Porter and J. F. Johnson, *Analytical Calorimetry*, Plenum, New York, 1968, p. 203.
15. H. J. V. Tyrrell, *Talanta*, **14**, 843 (1967).
16. F. Becker and R. Grundmann, *Z. Phys. Chem.* (New Series), **66**, 137 (1969).
17. F. J. Cioffi and S. T. Zenchelsky, *J. Phys. Chem.*, **67**, 357 (1963).
18. H. J. Keily and D. N. Hume, *Anal. Chem.*, **28**, 1294 (1956).
19. B. C. Tyson, Jr., W. H. McCurdy, Jr., and C. E. Bricker, *Anal. Chem.*, **33**, 1640 (1961).
20. P. W. Carr, *Thermochim. Acta*, **3**, 427 (1971/72).
21. D. Rosenthal, G. L. Jones, Jr., and R. Megargle, *Anal. Chim. Acta*, **53**, 141 (1971).
22. I. Sajó, *J. Therm. Anal.*, **1**, 349 (1969).
23. L. D. Hansen and E. A. Lewis, *Anal. Chem.*, **43**, 1393 (1971).
24. P. Marik-Korda, L. Buzási, and T. Cserfalvi, *Talanta*, **20**, 569 (1973).

CHAPTER

3

METHODS OF THERMOMETRIC TITRATION

In their review "Enthalpimetric Analysis," Jordan and Carr[1] distinguish two methods of titration: thermometric enthalpy titration (TET) and direct injection enthalpimetry (DIE). The first method consists of adding the titrant to the sample continuously or discontinuously; the second requires a fast and single addition of the titrant in excess. A further method also deserves mention: continuous-flow enthalpimetry,[2] in which the solutions flow continuously through a mixing vessel. This classification may not seem to be very penetrating, but it is close to analytical practice and allows a systematic development of theoretical problems based on the fundamental equation of thermometric titration discussed in Chapter 2.

3.1 THE TET METHOD

3.1.1 DESCRIPTION OF THE METHOD

The TET method has undoubtedly been the thermometric method used most frequently up to now. The theory of thermometric titration, presented in Chapter 2, is based on a discussion of thermograms obtained with this method.

The earliest form of the method is that described by Dutoit and Gröbet (Fig. 2.1), a manual method in which the titrand (solution A) is in a Dewar flask and the titrant (solution B) is added in small, equal quantities at equal intervals. The length of the time interval is adjusted to allow thermal equilibrium to be established and the temperature then to be read on an accurate liquid in glass thermometer.

Improved equipment became necessary to satisfy the growing interest in thermodynamic data of titration reactions and to reduce the amount of titrand with the concomitant necessity of measuring small changes of temperature.

The first requirement is a thermostated burette to measure these small changes. The Dewar vessel is also thermostated so that the burette and titration vessel have the same temperature. A stirrer is built into the Dewar flask to mix solutions A and B better. It is also advisable to provide it with an electric heater that enables the same equipment to be employed for the determination of the heat capacity C, specific heat c_B, and, if necessary, cooling constant κ [see (2.18) or (2.19)], quantities necessary for the evaluation of thermodynamic data.

The burette can be automatized, delivering the same quantity of titrant with every impulse of the control system. Useful devices for increasing the accuracy of temperature measurement are the multijunction thermocouple[3] and the thermistor as temperature sensor. Figure 3.1 shows equipment for thermometric titration that can be used for both continuous and noncontinuous (manual titration) addition of the reagent.[4]

The constant temperature θ of the environment in Fig. 3.1 is controlled with an accuracy of $\pm 0.002°C$ by a thermostat, which also controls the temperature of the burette (b). The reaction vessel is a Dewar flask (∅ 6 cm, depth 10 cm), completely immersed in the thermostated bath. The

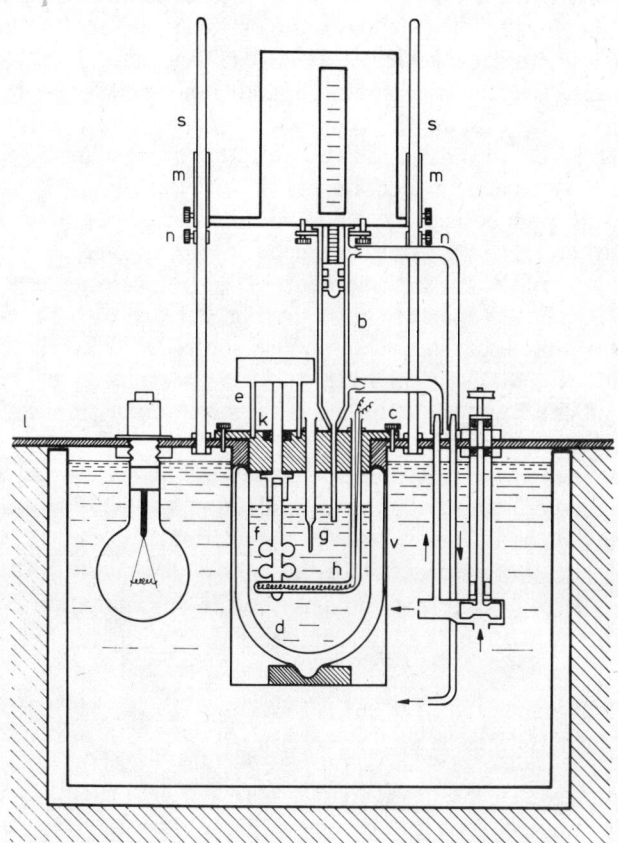

Fig. 3.1. Equipment for thermometric titration.[4] See text for explanation.

polypropylene lid (*c*) of the Dewar flask carries the motor of the stirrer. It is fitted with three supports (*e*) and a ball bearing (*k*) to diminish the friction of the stirrer axle. The glass stirrer is connected with the axle by means of a Teflon collar. The heater (*h*) is also fitted in the lid and so are the thermistor (*g*) and the tip of the burette. The titrant B is added from a motor-driven burette with remote control. The burette can be easily adjusted with the help of four guide collars (*m*) along the bars (*s*). Before the experiment the burette is let down to the guidance rings (*n*). The PVC lid (*l*) of the thermostat carries the control and measuring equipment necessary for its running. The container for the exchangeable reaction vessel (*d*) is also fitted to this lid.

A decisive step forward in the thermometric titration method was taken by Linde, Rogers, and Hume[5] with the construction of the first automatic titration calorimeter. In their calorimeter, the titrant is added continuously and at a constant speed dv/dt from the thermostated and motor-driven syringe to the titrand. The temperature is measured in the calorimeter vessel with a thermistor, which is placed in one arm of a Wheatstone bridge with direct current as source. The unbalance voltage of the bridge is registered as a measure of the temperature change by a recorder via a potentiometer. The automatic feed control of the paper in the recorder is adapted to the velocity of ejection of the burette. Figure 2.2 shows the results of using such equipment. The main problem is the quick mixing of solutions and an inertia-free measurement of temperature.

Technical details of the development of titration calorimeters are given in Chapter 9. The equipment shown in Fig. 3.1 can also be used as an automatic titration apparatus. The impulse-controlled delivery of titrant in fixed quantities and at equal time intervals is replaced by continuous delivery from the burette, controlled by a synchronous motor. The thermistor is suitably connected with a recorder.

The development of the automatic method for microanalysis,[6] as a differential method[7] and with digital indication,[8] constitutes further progress and has opened up new fields of application.

Thermograms are evaluated with the help of the methods discussed in Chapter 2. The following sections describe the determination of the necessary data.

3.1.2 DETERMINATION OF HEAT CAPACITY C, SPECIFIC HEAT c_B, AND COOLING CONSTANT κ

The determination of the heat capacity C (water equivalent) is a common calorimetric method. The calorimeter is filled with a given quantity of solution. After temperature balance has been reached, a preperiod *a*, of constant temperature gradient, begins. At time $t = t_0$ the heater of the calorimeter is switched on, preferably in combination with an electric stopwatch that is switched on by the same impulse. The heating period $\overline{AB} = b$ lasts until the temperature of the solution has risen by about 1°C. Then the heater and stopwatch are switched off simultaneously. An afterperiod c, about as long as the preperiod and also of constant temperature gradient, follows. The result is a T-t curve as depicted in Fig. 3.2.

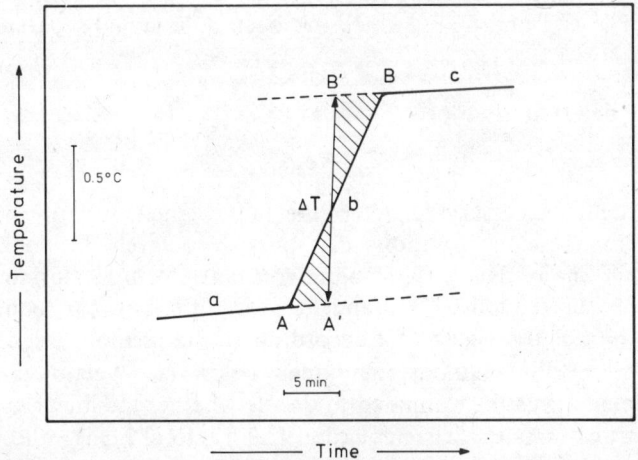

Fig. 3.2. Determination of the heat capacity of a filled reaction vessel. (*a*) preperiod; (*b*) heating period; (*c*) final period; (*A*) heater on; and (*B*) heater off.

The temperature rise $\Delta T \triangleq A'B'$ results if $\overline{A'B'}$ is drawn so that the two hatched triangles are of equal size. We then obtain

$$C = \gamma \frac{i^2 t}{R \Delta T} \tag{3.1}$$

where i[A] is the constant current passing through the heater of resistance R[Ω] during the heating period t[sec]. Here γ is a factor of dimension. With $\gamma = 1$ the result is C [J deg^{-1}]; with $\gamma = 0.2389$ the result is C with the dimension of cal deg^{-1}.

To determine the thermodynamic data of a titration reaction exactly, it is useful in some methods to measure the heat capacity of the calorimeter with different contents from C_0 to C_β. The result is a function according to (2.3). In the application of the section method (2.42a) and (2.42b) or (2.43) or the procedure proposed by Christensen et al. (see Section 2.3.7), the data C_α, C_β, or C_x can be deduced from such relationships.

If the calorimeter is filled with a solvent of known specific heat, C can be divided into the parts of the solvent and that of the calorimeter (water equivalent) in accordance with (1.4). The result is the calibration curve characterizing the calorimeter part

$$C_{\text{calorimeter}} = f(v)$$

It should be practically independent of v for a well-designed calorimeter.

The determination of the specific heat c_B depends on the same measurement principle. The calorimeter is filled with the solution B. The separation of C_{sol} and C_{calor}, which has been determined separately,

$$C_{\text{tot}} = C_{\text{sol}} + C_{\text{calor}} \tag{3.2a}$$

allows the determination of c_B

$$C_{\text{sol}} = c_B V_{\text{sol}} \tag{3.2b}$$

It is frequently necessary to determine these values, notably because c_B values of nonaqueous solutions can only very rarely be found in the literature. A knowledge of the cooling constant κ of the calorimeter is not essential for the section or the tangent procedure, but it is required for a determination of the heat effect according to the method based on (2.64) and several of the methods mentioned below (see Section 3.1.4). The magnitude of κ must be estimated to decide whether the linear approximation that transforms the basic equation (2.20) into (2.21) is valid or not.

The constant κ is determined by plotting a cooling curve. The calorimeter is filled with a content of heat capacity about C_α and, after thermal equilibrium is attained, the calorimeter content is heated by about 2°C. The temperature T is then measured at regular intervals (ca. 5 min) for about 2 h stirring uniformly, to yield it as a function of time. The thermal balance of cooling is defined as

$$C\frac{dT}{dt} = -C\kappa(T-\theta) + w_0 \tag{3.3}$$

The application of Guggenheim's method of evaluation for the first-order rate equation results in a linear function for the cooling curve[9]

$$\log(T-T') = -0.4343\kappa t + \log(e^{-\kappa\tau}-1)\left(\frac{w_0}{C\kappa} - T_0 + \theta\right) \tag{3.4}$$

This is done by grouping the measured values into two series so that each corresponding pair of values is separated by the same interval τ, that is, if one series (over about 60 min) has been measured as $T(t_0), T(t_1), \ldots, T(t_n)$, the second series will be measured as $T'(t_0+\tau), T'(t_1+\tau), \ldots, T'(t_n+\tau)$. The slope of the linear function equation (3.4) gives κ with sufficient accuracy ($\pm 5\%$).

3.1.3 CONTINUOUS OR DISCONTINUOUS ADDITION OF TITRANT

It is an interesting question whether the "sharpest" indication of a titration end point can be obtained with the help of continuous or discontinuous addition of the titrant. An inspection of the data published so far does not give a definite answer. Older investigations, using the discontinuous method, were conducted with much simpler equipment than that necessary for the more recent continuous titration with thermostated burette, thermistor, and so forth. On the basis of personal experience it can be stated that with an experimental setup such as that shown in Fig. 3.1 the end point is certainly indicated as sharply in discontinuous work as in the continuous method. Any time lag in the temperature can easily be recognized with the discontinuous method, whether is due to the measurement equipment or to kinetic inhibition of the reaction. Brown, Issa, and Sinclair[10] have experimented with a similar apparatus. Up to 90% of the titration reagent is added quickly, after which it is added in steps of 0.1 ml at equal time intervals of about 10 to 15 sec. The resulting diagram is plotted by a recorder, shown in Fig. 3.3. The points of intersection of the horizontal of temperature with the time line of the recorder are connected and extrapolated to give the end point.

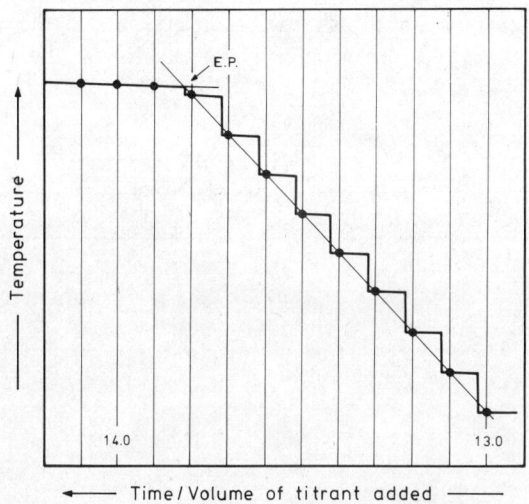

Fig. 3.3. Thermogram of a titration with discontinuous adding of titrant near the end point (EP). The vertical lines are the cross lines of the chart, and the heavy stepped line is the chart pentrace.[10]

The authors claim two advantages for the procedure:

1. The extrapolation has a margin of error of only 0.01 ml and is thus more accurate than that obtained from the curve recorded with continuous addition of the titrant.
2. The calculation of the titer is facilitated.

The results obtained by these authors (see Table 3.1) are among the most exact published so far.

TABLE 3.1. Accuracy of End-Point Determination Using a Titrant Added Discontinuously near the End Point[10]

Titrand	Titrant	Number of Titrations	Titer ml	Standard deviation ml	%
0.025 M I^-	0.50 M Ag^+	10	4.98	0.006	0.13
0.0375 M Cl^-	0.50 M Ag^+	10	7.50	0.019	0.25
0.025 M Ce(IV)	0.50 M Fe(II)	9	5.00	0.011	0.21
0.025 NH_4^+	0.27 M OCl^-	10	13.75	0.013	0.09
0.0125 M I^-	0.27 M OCl^-	7	13.71	0.020	0.15
0.0125 M $S_2O_3^{2-}$	0.07 M OCl^-	5	18.15	0.011	0.06

Basic investigations have also been carried out on the discontinuous addition of titrant and its effect on the results of the titration using potentiometric methods.[11,12]

TABLE 3.2. Accuracy of End-Point Determination Using Continuously Added Titrant

Titrand	Titrant		Number of Titrations	Amount Added	Amount Found meq	Standard Deviation %	Ref.
OH^-	H^+	1.008 N	6	9.804	9.804	0.11	5
Ag^+	Cl^-	1.008 N	8	5.000	4.98	0.25	5
OH^-	H^+	1.013 N	6	1.150	1.140	0.10	7
Ag^+	Cl^-	0.9698 N	10	0.2987	0.2988	0.5	7
Fe(II)	Ce(IV)	0.3483 N	3	0.2123	0.2118	0.4	7
OH^- $M/60$	H^+		10	–	–	0.6	8
Ag^+ $M/60$	Cl^-		10	–	–	0.5	8

Table 3.2 contains results obtained in automatic titration with continuously added titrant. The advantage of continuously adding titrant is that the complete thermogram is plotted directly by the recorder, making direct evaluation of the process of titration possible. The methods of evaluation of diagrams require a titration curve to be plotted and are facilitated by the continuous plotting procedure; for the initial slope method, a titration curve is, in fact, imperative. The continuous addition of titrant is not a necessary feature of entropy titrations. Several scientists[13,14] use discontinuous addition and cooling down to the initial temperature after each addition of titrant.[15,16] In our own laboratory we use measuring equipment (see Chapter 9) that regulates the addition in steps of preset duration, printing the data obtained or storing them for further processing.

The DIE method, which is described in the next section, requires a recorder. Multiple method equipment, therefore, always possesses the necessary outfit to permit use of the continuous TET method.

Both methods, continuous as well as discontinuous addition, are still justified even at today's level of technological achievement; the choice of which to use is not a matter of principle.

3.1.4 RECENT DEVELOPMENTS OF TET TECHNOLOGY

Titration calorimeters of the type commonly used do not take into account the influence of the gas phase above the solution in the reaction vessel. The gas phase space is mainly responsible for two errors: measurements on substances sensitive to air or moisture are falsified; and solutions with a high vapor pressure cause errors through evaporation. Our discussion, based on practical experience of thermometric titration, has so far assumed that the gas phase of the titration vessel should be kept so small that only a small amount of titrant B can be added. Such an arrangement is not, however, always possible. Recent developments aim at a volume of addition of up to 100% and more of the titrand. Two different principles can be recognized in this development. In some methods the volume of the calorimeter vessel is varied as the titrant B is added[17]; in a second group, the volume V_0 of the reaction vessel is kept constant and the reacted mixture of the two solutions is run off after thermal equilibrium has been attained with the same velocity v'_0 as that with which solution B is added.[18] The evaluation of thermograms plotted according to this overflow method requires a small change in the basic equation (2.18). The loss of thermal power due to the overflow

$$w_{ov} = -\frac{v'_0}{V_0} C(T-\theta) \tag{3.5}$$

must be included in the heat balance of (2.18) and may be interpreted as an increase of the cooling constant by the amount v_0'/V_0. The heat capacity C varies with changing calorimeter content; only for dilute solutions $C = C_0$. Instead of (2.18), the following basic equation must be used:[19]

$$\frac{d}{dt}[C(T-\theta)] = w_R(t) + w_v(t) + w_0 - C\left(\kappa + \frac{v_0'}{V_0}\right)(T-\theta) \quad (3.6)$$

Reference 19 also includes a discussion of the possible application of further calorimetric methods, namely, a quasiisothermal flow method and an isothermal method with controlled compensation of thermal power.

3.1.5 THERMOMETRIC TITRATION AS A METHOD FOR DETERMINING ENTHALPY OF MIXING

The possibility of adding the titrant up to a ratio 1:1 or more has opened up a new and promising field of application for thermometric titration as a method of investigating the integral enthalpy of mixing, H^M, of binary or multicomponent systems.[20] The thermometric method is clearly superior to the classical calorimetric method in this domain because the latter can plot the function $H^M = f(N_A)$ only in a time-consuming manner from individual points that must be determined in separate experiments.

With the thermometric method, the function H^M is obtained over the entire range of the molar fraction $0 \leq N_A \leq 1$ of the components A and B with two experiments only. Component B is first added to the sample A up to the ratio 1:1; in the second, B is the sample and A is added up to the same ratio. The thermograms are of the type in Fig. 2.5. Preperiod and afterperiod are given by (2.25a) and (2.25c). The titration period is determined by the equation

$$[C(T-\theta) - C_0(T_0-\theta)](1 + \tfrac{1}{2}\kappa t)$$
$$= \int_0^t w_M(t)\,dt + w_0 t - C_0 \kappa t(T_0-\theta) \quad (3.7)$$

where

$$Q_M(t) = \int_0^t w_M(t)\,dt \quad (3.8)$$

is the integral heat effect for the process of mixing between $t=0$ and t; the other symbols have their usual meaning. Determination of $Q_M(t)$ or $w_M(t)$ can be made either by the section method (Section 2.3.4), the tangent method (Section 2.3.5), or any other convenient procedure. With these values we obtain the integral enthalpy of mixing

$$H^M(t) = -\frac{Q_M(t)}{n_A^0 + n_B(t)} \tag{3.9}$$

and the partial molar enthalpy of mixing

$$H_B^M(t) = -\left(\frac{\partial Q_M}{\partial n_B}\right)_{n_A} = -w_M(t)\frac{M_B}{v_0'\rho_B} \tag{3.10}$$

when component B is added to the sample A; v_0' is the velocity (liters sec^{-1}) of adding pure B from the burette, M_B the molar mass (kg) of B, and ρ_B (kg liter^{-1}) its density. The time coordinate is transformed into the molar fraction with the help of the equation

$$n_B(t) = \frac{v_0'\rho_B}{M_B}t \tag{3.11}$$

As an example, Fig. 3.4 shows the result of the titration of acetone with chloroform.[18] This investigation was conducted with the overflow method (Section 3.1.4) in connection with an isothermal calorimeter; (3.7) must be transformed in the usual way to take into account these experimental conditions. The efficiency of the method can be seen from Fig. 3.5; with a few experiments (dotted lines) a complete diagram for the integral heat of

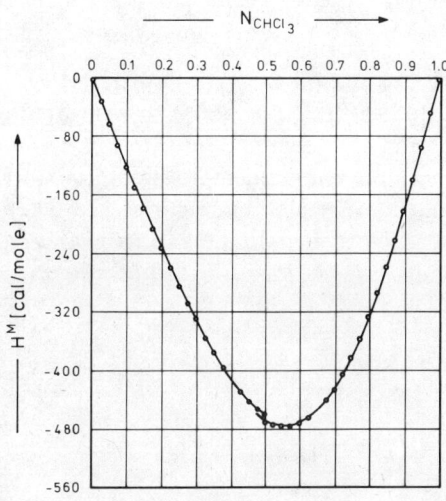

Fig. 3.4. Heat of mixing $H^M = f(N_{CHCl_3})$ of the system acetone-chloroform.[18] The function H^M is obtained with two experiments: addition of chloroform to acetone up to the ratio 1:1 and reverse titration up to the same ratio.

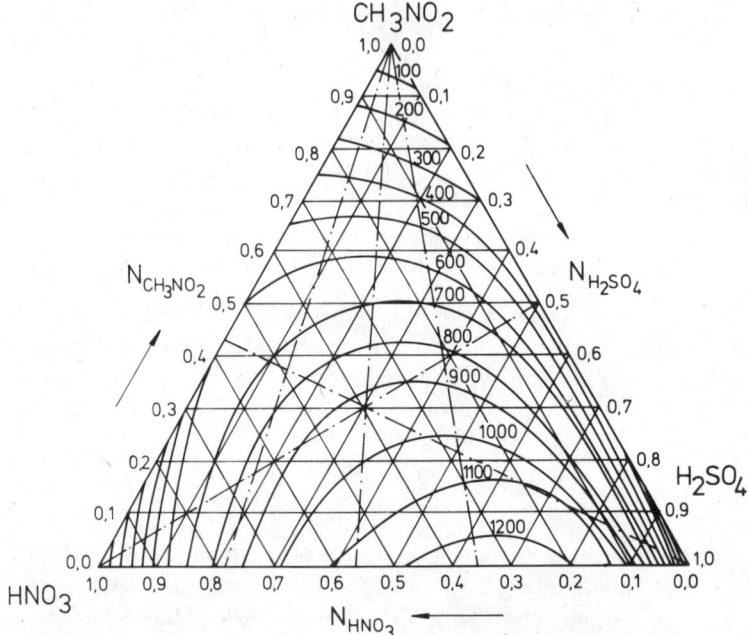

Fig. 3.5. Integral enthalpy of mixing H^M of the ternary system H_2SO_4—HNO_3—CH_3NO_2, determined by thermometric titration technique.[20] The diagram contains the isenthalpic lines of the system with the corresponding enthalpy values. The total information is obtained by titrating only five binary mixtures with the third pure component (–·–·– paths on which the experiments were conducted).

mixing of a ternary mixture (H_2SO_4–HNO_3–CH_3NO_2) can be evaluated.[20] In comprehensive studies Becker et al.[18–26] investigated enthalpies of mixing for various systems by thermometric titration methods; they interpret their results by a thermodynamic theory of liquid mixtures that is based on an equilibrium model. Molecular interactions are taken into account by complex formation between a molecule and its nearest neighbors. Theoretical treatment of measuring values is based on the principles of multistep reactions (Section 8.2.2).

3.2 THE DIE METHOD

3.2.1 DESCRIPTION OF THE METHOD

The DIE method was introduced in 1964 by Wasilewski, Pei, and Jordan;[27] it is the reapplication of an old calorimetric method. A volume of titrant B is added very quickly to the titrand A. Titrant B has such a high concentration that the reaction [cf. (2.8)]

$$x_A A + x_B B \rightleftarrows z_P P \tag{3.12}$$

takes place completely and quickly (in this case within 10^{-2} sec), leaving a surplus of B. Registering the temperature as a function of time yields an injection enthalpogram (Fig. 3.6). Instead of the temperature jump ΔT, the unbalance voltage ΔE in the thermistor bridge can be measured and used directly for determination of the titer of A when the bridge has been appropriately calibrated.

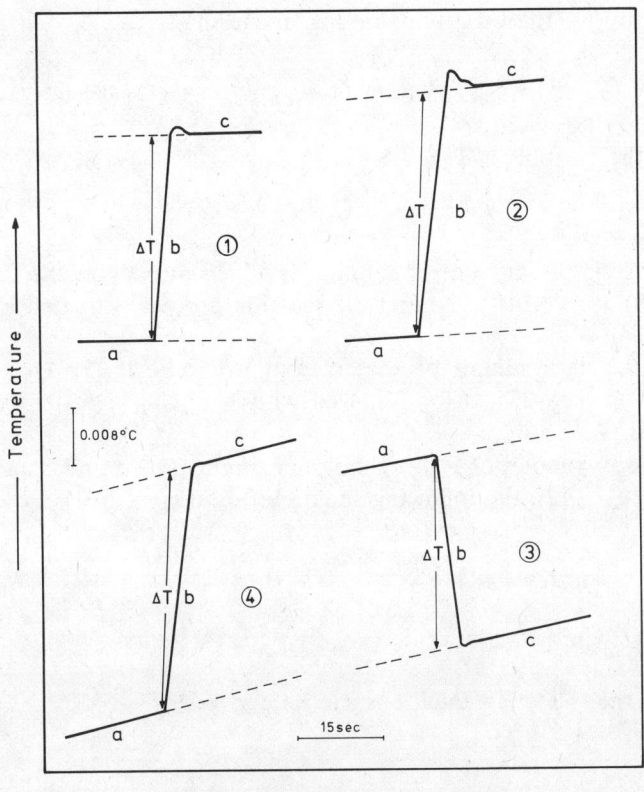

Fig. 3.6. Typical injection enthalpograms.[27] (a) preperiod; (b) titration period; and (c) final period. Injection of 300 μl 1 M reagent into 25 ml volume of unknown solution. (1) 83.1 μmol of HCl plus an excess of NaOH, $\Delta H_R = -13.5$ kcal mol^{-1}; (2) 101 μmol of H_3BO_3 plus an excess of NaOH, $\Delta H_R = -10.6$ kcal mol^{-1}; (3) 100 μmol of Mg^{2+} plus an excess of EDTA, $\Delta H_R = +5.5$ kcal mol^{-1}; and (4) 104 μmol of Pb^{2+} plus an excess of EDTA, $\Delta H_R = -12.8$ kcal mol.$^{-1}$

3.2.2 THEORETICAL BASIS OF THE METHOD

An injection enthalpogram, such as that in Fig. 3.6, consists of three parts: the preperiod, the titration period, and the afterperiod.

The data quoted from the original paper[27] describing the experimental setup show that the reaction is so fast that the heat of stirring and heat exchange with the environment do not play any noticeable part. The resolution on the time axis of the T-t diagram is not sufficiently high to produce a break in the curve indicating the end of the reaction and the beginning of the phase of dilution. The thermogram is characterized by the system of (2.19) with the following simplifications for the titration period: $\kappa t = 0, w_0 t = 0$.

The result for the titration period is, according to (2.21),

$$C(T-\theta) - C_0(T_0-\theta) = \int_0^t (w_R + w_v) \, dt = Q_R(t) + Q_v(t) \quad (3.13)$$

or, using the symbols of Fig. 3.6,

$$C_\beta \Delta T = Q_R + Q_v - c_B \Delta v (T_0 - \theta) \quad (3.14)$$

in which Δv is the amount of solution B added at once at the time $t=0$. Both (3.13) and (3.14) show that the reaction process is treated as strictly adiabatic.

For the determination of the integral heat effect, $Q = Q_R + Q_v$, the discussion in Ref. 27 can be followed without taking over its hypothesis that $Q_v = 0$.

If we add n_B^0 moles of B to n_A^0 moles of titrand A, we find, in a titration according to (3.12), the following situation $[n_B^0 > (x_B/x_A)n_A^0]$:

$$[n_A^0 A + n_B^0 B] \rightarrow \left[\frac{z_P}{x_A} n_A^0 P + \left(n_B^0 - \frac{x_B}{x_A} n_A^0 \right) B \right] \quad (3.15)$$

$$\text{initial state} \qquad \text{final state}$$

From this the following equations can be derived:

$$Q_R = -\frac{z_P}{x_A} n_A^0 \Delta H_R = k n_A^0 \quad (3.16)$$

and

$$Q_v = -n_B^0 H_v \quad (3.17)$$

if H_v' is neglected. By the use of (3.16) and (3.17),

$$\Delta T = \frac{1}{C_\beta} (k n_A^0 + k') \quad (3.18)$$

can be derived from (3.14), in which $k' = -n_B^0 H_v - c_B \Delta v(T_0 - \theta)$. Two titrands of known concentration can be used for the determination of the coefficients k/C_β and k'/C_β; without changing the experimental conditions, titrations of an unknown n_A^0 can be conducted in which ΔH_R is determined simultaneously.

The procedure is especially simple and scarcely susceptible to interference because the reaction process is almost adiabatic; it is also fast. Figure 3.6 shows the duration of the reaction to be less than 1 min. If the accuracy of calibration and the constancy of coefficients k and k' are taken into account, the titer can be determined to within 2 to 3%.

It is noteworthy that the titration reaction of (3.12) must be fast, but that it need not be complete. It is sufficient for the reaction to be almost completed when an excess of B is added.

3.2.3 APPLICATIONS OF THE METHOD

The DIE method has a wide field of application; it offers a direct and profitable approach to microanalysis,[6] for which Table 3.3 gives characteristic results.

TABLE 3.3. Precision and Accuracy in Microcalorimetric Titrations and Microinjection Enthalpimetry[6]

Reaction		Applicable Range Investigated			Precision	
Unknown	Reagent	Concentration mmol liter^{-1}	Total Sample μmol liter^{-1}	Number of Replicates	Standard Deviation %	Accuracy %
Strong acid	Strong base	1–20	2–40	6	±1	1–2
Pb^{2+}	EDTA	2–20	4–40	5	±1	1–2
Sulfanilamide	NaNO$_2$	1–100	2–200	5	±1	1

The work of Sajó et al. impressively demonstrates the significance of the method for industrial testing laboratories. These authors have developed methods for the control of various products by direct-injection procedure and with direct percentage-reading equipment (see Chapter 9). Examples are the quick analysis of blast furnace and Siemens–Martin slags;[28,29] pig iron and steels;[29–31] plating solutions;[32] and cements, clays, magnesites, dolomites, and industrial silicates.[33–36] Details of this work and of further developments are given in the following sections.

Another example of industrial application is the use of DIE in the routine determination of the hydroxyl value of alkyl phenols.[37] Early DIE techniques are applied in the determination of benzene in the presence of cyclohexane[38,39], in a quick moisture method for coal,[40] and for a rapid estimation of the strength of sulfuric acid.[41] Further applications of the DIE methods are given in the following chapters.

3.3 CONTINUOUS-FLOW ENTHALPIMETRY

3.3.1 DESCRIPTION OF THE METHOD

The principle of measurement[2] consists in passing solutions A (a, mol liter^{-1}) and B (b, mol liter^{-1}) through the measuring cell (Fig. 3.7) at constant velocities. Quick mixing of the reactants and their reaction to produce P take place in the mixing chamber M. The temperatures T_A and T_B of the solutions entering, and T_P of the final product leaving, the measuring cell are measured with the help of thermistors T_a, T_b, and T_p. Determined is the titer of solution A, of which there is a stoichiometrically smaller amount than of B.

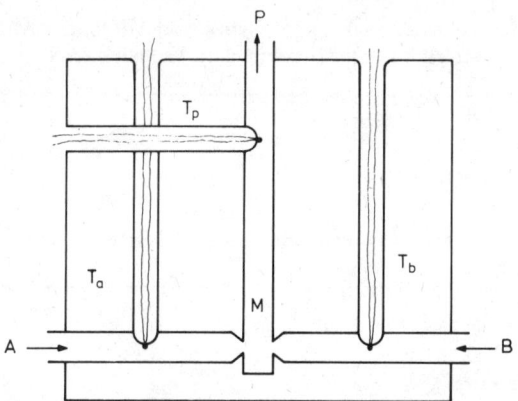

Fig. 3.7. Diagram of a reaction vessel for continuous-flow enthalpimetry.

3.3.2 THEORETICAL BASIS OF THE METHOD[2]

To simplify the derivation of a suitable equation, we assume the heat capacities of the initial reactants A and B and that of the final product P to be constant and equal: $C_A = C_B = C_P = C$. Such an assumption does not lead to significant errors for aqueous solutions of concentration less than 1 mol liter^{-1}. The flow rate of A is termed x, that of B is Rx. Then R is the ratio of the flow rates. On the basis of the assumptions made above, $xa < xRb$. As always ΔH_R is the enthalpy of reaction.

The heat balance

heat entering the system = heat leaving the system

can be written as

$$xa\Delta H_R + xCT_A + RxCT_B = (R+1)xCT_P \qquad (3.19)$$

for the resulting stationary state if we neglect the heat exchange with the environment, since the unreacted solutions transport a heat amount per second equal to $Cx(T_A + RT_B)$ into the measuring cell and a heat amount $(R+1)xCT_P$ leaves with the final product. The heat produced during the reaction is $xa\Delta H_R$. Equation (3.19) can be rewritten as

$$\frac{a\Delta H_R}{C} = T_P(R+1) - T_A - RT_B \qquad (3.20)$$

Equation (3.20) no longer contains the flow rate x. The titration depends solely on the variables in (3.20). When these are kept constant, we obtain (3.20) as a linear function $a = f(T_P, T_A, T_B)$.

There are several variants of the continuous-flow method for different domains of application.

1. $R = 1$ $\quad a = \left(T_P - \dfrac{T_A + T_B}{2}\right)\dfrac{2C}{\Delta H_R}\quad$ used in the concentration range $0.01M < a < 1M$

2. $R < 0.01$ $\quad a = (T_P - T_A)\dfrac{C}{\Delta H_R},\quad$ used in the concentration range $a < 0.01M$

3. $R > 100$ $\quad a = (T_P - T_B)\dfrac{CR}{\Delta H_R}\quad$ used in the range of high concentration.

Among other applications is one with a discontinuous addition of the titrant.[2]

3.3.3 APPLICATION OF THE METHOD

Table 3.4 contains results obtained with the help of an experimental setup as described previously[2]

TABLE 3.4. Comparison of Observed and Nominal Values[2]
(Reagent 0.6 M NaOH)

Sample	Results			
Nominal Values:	0.050	0.100	0.300	0.500
HCl	0.051	0.103	0.303	0.500
HNO$_3$	0.050	0.100	0.300	0.500
H$_3$BO$_3$	0.049	0.099	0.302	0.500
CH$_3$COOH	0.050	0.098	0.305	0.500

After calibration of the apparatus through a reaction of known reaction enthalpy, for example, the neutralization reaction ($\Delta H_R = -13.5$ kcal mol^{-1}), it is possible to determine unknown enthalpies of reaction if the heats of dilution are taken into account.

The accuracy of the continuous-flow method is satisfactory. The mixing of reactants in the absence of a gaseous phase is one of the advantages of this method; this is of importance in experiments with volatile liquids. A special field of application is the use as on-line process control.

Taubinger[42] describes a continuous-flow enthalpimetric analyzer for the direct determination of free acrylonitrile in aqueous copolymer latices. Crompton and Cope[43] apply this method to measure total organo-aluminum-reactable impurities (water, dissolved oxygen, low-molecular-weight alcohols) in the control of hydrocarbons.

3.3.4 FLOW MICROCALORIMETRY

Flow microcalorimetry is a new analytical tool of special interest in investigations into complex biochemical or microbiological processes.[44] Since there is no gaseous phase in the measuring system, condensation and evaporation effects cannot affect the result. Surface adsorption effects can be neglected if a steady liquid flow is allowed to continue until wall reactions have occurred.

A flow microreaction calorimeter has been designed (for construction details see Chapter 9) and tested by Monk and Wadsö.[44,45] For a normal operational procedure the reaction time must be short compared to the retention time for the liquid in the measuring cell (in the present case, about 5 min); for slow reactions a *stopped-flow technique* can be applied, where a pulse of the reactants is pumped into the flow cell and kept there until the reaction is completed. Alternatively, *flow techniques on premixed reaction systems* can be used for slow reactions. In this type of experiment only a part of the reaction heat is measured, as the reaction is started outside the measuring cell. Quantitative evaluation is achieved by comparison of the calorimeter curves with those obtained in calibration experiments. This method has been used to determine enzyme activities for glucose oxidase, cholinesterase, alkaline phosphatase, lactic acid dehydrogenase, and for ATPase activity in a tissue homogenate.[44] The method is also demonstrated in the determination of glucose concentration;[45] its usefulness in pesticide research and related areas has been explored.[46] Beezer and Stubbs have also applied flow microcalorimetry to the determination of organophosphorus pesticides via their inhibition of cholinesterase.[47]

3.4 SPECIAL METHODS FOR INCREASING THE ACCURACY OF TITRATION

The lower limit c_x (mol liter^{-1}) of the titrand concentration still attainable at a prescribed error of $x\%$ depends on the features of the apparatus and the properties of the titrand/titrant system. If the latter furnishes a heat of reaction of ΔH_R (kcal mol^{-1}) per mole of titrand, of which V (liters) of concentration c_x is to be titrated, and the error of temperature measurement is $\Delta\epsilon$ (K), then c_x may be estimated according to

$$c_x \approx \frac{100 \Delta\epsilon}{x \Delta H_R} \frac{C}{V} \quad (3.21)$$

where C is the heat capacity (kcal deg^{-1}) of the system.[30] For aqueous solutions with $C/V \approx 1$ and $\Delta H_R \approx 10$ kcal mol^{-1}, (3.21) leads to $c_x \approx 10\Delta\epsilon/x$, that is, with $\Delta\epsilon = 10^{-3}$ and a prescribed error of $x = 1\%$, $c_x \approx 10^{-2}$ mol liter^{-1}. Equation (3.21) reveals two possible ways of lowering the limit c_x. The first is through improvement of the technical aspect, which is accompanied by a reduction of $\Delta\epsilon$; the other is to find reactions with the highest possible ΔH_R. The technical ways and means are discussed in Chapter 9. Among "chemical possibilities," Sajó and Sipos[30] suggest the following combinations:

3.4.1 COMBINATION WITH A SECONDARY REACTION OF HIGH HEAT CHANGE

EXAMPLE. In the determination of sulfur in steel, cast iron, or slags, the element is liberated as hydrogen sulfide during solution of the sample in hydrochloric acid; this is converted to sodium sulfide by means of sodium hydroxide. The sulfide is oxidized with potassium permanganate in alkaline solution to sodium sulfate, so that the heat of oxidation and also the fourfold heat of neutralization per atom of sulfur are developed

$$Na_2S + 8KMnO_4 + 8KOH \rightleftharpoons Na_2SO_4 + 8K_2MnO_4 + 4H_2O; \Delta H_R \approx 100 \text{ kcal mol}^{-1}$$

3.4.2 INDIRECT THERMOMETRIC ANALYSIS

An indirect thermometric analysis has two stages. In the first stage, each molecule of the titrand reacts with several molecules of titrant, for example, to yield a precipitate. This primary step has a heat of reaction of a few kilocalories per mole. After its conclusion, the precipitate is filtered and subjected to a further chemical change.

EXAMPLE. Phosphorus in steel is oxidized to phosphoric acid on dissolving the sample in nitric acid. This is precipitated as $(NH_4)_3PO_4 \cdot 12MoO_3$ with ammonium molybdate. The heat of neutralization of 12 mol molybdic acid anhydride per atom of phosphorus is yielded on titrating the precipitate with NaOH, instead of the heat of neutralization of only the 1 mol of phosphoric acid that would be furnished in direct titration of this acid with the NaOH.

3.4.3 KINETIC TITRATION

The heat of reaction, normally only a few kilocalories per mole, is amplified in catalyzed reactions to an extent dependent on the concentration of the catalyst. This is because 1 mol of the catalyst develops the molar heat of reaction for each mole that is combined or separated through the catalyst during the period of measurement. A distinct change of temperature is then obtained in very dilute solution, in which no measurable change would be observed with other methods. In this case, however, the concentration of the component studied is obtained not from the temperature change of the solution, but from the rate of change of temperature,[30,48] dT/dt, which is proportional to the rate of the catalytic reaction, dc/dt,

$$\frac{dT}{dt} = k\frac{dc}{dt} \qquad (3.22)$$

where k is a proportionality constant.

At low concentrations, dc/dt is proportional to the concentration of the catalyst, and hence so is dT/dt. The kinetic titration method has enabled thermometric analysis to be extended to solutions[30] of 10^{-6} to 10^{-7} mol liter^{-1}.

EXAMPLE. Cobalt impurities in nickel metal can be determined by means of kinetic titration, since cobalt ions catalyze the decomposition of hydrogen peroxide in alkaline solution. The change of temperature of the sample solution is proportional to the concentration of cobalt ions. The addition of tartaric acid prevents precipitation of the hydroxides of cobalt and of any other metals possibly present.

Sajó[49] developed kinetic titration as a differential method in which two solutions, identical except for the addition of catalyst to one, are compared. The reference solution without catalyst is then titrated with a dilute solution of known catalyst concentration until the catalyst has the same concentration in both solutions, as shown by the identity of the rates of reaction.

Another thermometric kinetic analytical method based on (3.22), in which the reaction utilized is the solution process of a metal, has been used by Sajó and Sipos for direct determination of impurities in metals.[30]

EXAMPLE. Aluminum metal is dissolved in hydrochloric acid at a rate that, for a number of its natural impurities, increases in proportion to their concentrations.

3.5 METHODS WITH AMPLIFIED END-POINT INDICATION

3.5.1 CATALYTIC THERMOMETRIC TITRATION

In this procedure the break constituting the end point is intensified through the initiation of catalysis by excess titrant of a reaction of the solvent or of a component added to the solvent. This important procedure is employed in various forms and is dealt with later in Sections 5.5.1 (titration in nonaqueous media), 6.2.10 (precipitation titration), and 8.3.3 (complexometric titration).

3.5.2 INTENSIFICATION OF END POINT THROUGH HIGH HEAT OF DILUTION

A similarly functioning procedure is based on the large heats of dilution of certain titrant solutions. It is associated with nonaqueous solvents and is accordingly discussed in Section 5.5.2.

REFERENCES

1. J. Jordan and P. W. Carr, in R. S. Porter and J. F. Johnson, *Analytical Calorimetry*, Plenum, New York, 1970, p. 203.
2. P. T. Priestley, W. S. Sebborn, and R. F. W. Selman, *Analyst*, **90**, 589 (1965).
3. R. H. Müller, *Ind. Eng. Chem.*, Anal. Ed., **13**, 667 (1941).
4. J. Barthel and N. G. Schmahl, *Z. Anal. Chem.*, **207**, 81 (1965).
5. H. W. Linde, L. B. Rogers, and D. N. Hume, *Anal. Chem.*, **25**, 404 (1953).
6. J. Jordan, R. A. Henry, and J. C. Wasilewski, *Microchem. J.*, **10**, 260 (1966).
7. B. C. Tyson, Jr., W. H. McCurdy, Jr., and C. E. Bricker, *Anal. Chem.*, **33**, 1640 (1961).
8. P. T. Priestley, *Analyst*, **88**, 194 (1963).

9. J. Barthel, F. Becker, and N. G. Schmahl, *Z. Phys. Chem.* (New Series), **29**, 58 (1961).
10. M. W. Brown, K. Issa, and A. G. Sinclair, *Analyst*, **94**, 234 (1969).
11. A. Johansson, *Analyst*, **95**, 535 (1970).
12. A. Johansson and L. Pehrsson, *Analyst*, **95**, 652 (1970).
13. P. Gerding, I. Leden, and S. Sunner, *Acta Chem. Scand.*, **17**, 2190 (1963).
14. R. Arnek, *Ark. Kemi*, **24**, 531 (1965).
15. K. Schlyter and L. G. Sillén, *Acta Chem. Scand.*, **13**, 385 (1959).
16. I. Danielsson, B. Nelander, S. Sunner, and I. Wadsö, *Acta Chem. Scand.*, **18**, 995 (1964).
17. C. G. Savini, D. R. Winterhalter, L. H. Kovach, and H. C. van Ness, *J. Chem. Eng. Data*, **11**, 40 (1966).
18. F. Becker and M. Kiefer, *Z. Naturforsch.*, **24a**, 7 (1969).
19. F. Becker, *Chem. Ing. Tech.*, **41**, 1060 (1969).
20. F. Becker, N. G. Schmahl, and H. D. Pflug, *Z. Phys. Chem.* (New Series), **39**, 306 (1963).
21. F. Becker, *Chem. Ing. Tech.*, **41**, 1105 (1969).
22. F. Becker, H. D. Pflug, and M. Kiefer, *Z. Naturforsch.*, **23a**, 1805 (1968).
23. F. Becker, E. W. Fries, M. Kiefer, and H. D. Pflug, *Z. Naturforsch.*, **25a**, 677 (1970).
24. F. Becker and M. Kiefer, *Z. Naturforsch.*, **26a**, 1040 (1971).
25. F. Becker, M. Kiefer, and P. Rhensius, *Z. Naturforsch.*, **27a**, 1611 (1972).
26. F. Becker, M. Kiefer, and H. Koukol, *Z. Phys. Chem.* (New Series), **80**, 29 (1972).
27. J. C. Wasilewski, P. T.-S. Pei, and J. Jordan, *Anal. Chem.*, **36**, 2131 (1964).
28. I. Sajó and B. Sipos, *Z. Anal. Chem.*, **222**, 23 (1966).
29. I. Sajó and J. Ujvári, *Z. Anal. Chem.*, **202**, 177 (1964).
30. I. Sajó and B. Sipos, *Mikrochim. Acta*, **1967**, 248.
31. I. Sajó, *Z. Anal. Chem.*, **242**, 165 (1968).
32. I. Sajó and B. Sipos, *Talanta*, **14**, 203 (1967).
33. I. Sajó and B. Sipos, *Radex-Rundsch.*, **1968**, 178.
34. I. Sajó and B. Sipos, *Zem.-Kalk-Gips*, **21**, 32 (1968).
35. I. Sajó and B. Sipos, *Tonind. Z. Keram. Rundsch.*, **92**, 88 (1968).
36. I. Sajó and B. Sipos, *Bull. Soc. Fr. Céram.*, **85**, 3 (1969).
37. F. L. Snelson, W. R. Ellis, and J. Vilkauls, *Analyst*, **92**, 264 (1967).
38. B. B. Corson and L. J. Brady, *Ind. Eng. Chem.*, Anal. Ed., **14**, 531 (1942).
39. R. L. Bishop and E. L. Wallace, *Ind. Eng. Chem.*, Anal. Ed., **15**, 563 (1943).
40. V. R. Gray and P. F. Whelan, *Chem. Ind.*, **1955**, 126.
41. H. D. Richmond and J. E. Merreywether, *Analyst*, **42**, 273 (1917).
42. R. P. Taubinger, *Analyst*, **94**, 634 (1969).
43. T. R. Crompton and B. Cope, *Anal. Chem.*, **40**, 274 (1968).
44. P. Monk and I. Wadsö, *Acta Chem. Scand.*, **23**, 29 (1969).
45. P. Monk and I. Wadsö, *Acta Chem. Scand.*, **22**, 1842 (1968).
46. J. Koničková and I. Wadsö, *Acta Chem. Scand.*, **25**, 2360 (1971).
47. A. E. Beezer and C. D. Stubbs, *Talanta*, **20**, 27 (1973).
48. I. Sajó, *Vasipari Kut. Intéz. Közl.*, **2**, 687 (1965); quoted after Ref. 28.
49. I. Sajó, *Talanta*, **15**, 578 (1968).

CHAPTER

4

ACID-BASE TITRATIONS IN AQUEOUS SOLUTIONS

Acid-base titrations were, not unexpectedly, among the first thermometric titrations.[1,2] Bell and Cowell[1] used the thermometric indication of end points to prepare neutral solutions of ammonium citrate; this method successfully superseded the uncertain method using an alcoholic solution of corallin as indicator. This first investigation was conducted in the context of agricultural chemistry and is the predecessor of later thermometric titration. Dutoit and Grobet[2] introduced thermometric titration in their article "Sur une nouvelle méthode de volumétrie physicochimique appliquée à quelques problèmes de Chimie minérale." Among their examples of end-point indication in this "new method of analysis" were titrations of sulfuric acid and phosphoric acid with caustic soda solution.

The simplest acid-base titrations are those of strong monoacids with strong monobases in aqueous solution. This type of reaction must be discussed in detail because it is frequently used as a standard for calibration.

4.1 THE HEAT EFFECT OF STANDARD REACTIONS

The neutralization reaction

$$H^+(aq) + OH^-(aq) \rightleftharpoons H_2O(l) \tag{4.1}$$

has been frequently investigated. The heat effect of this reaction is the heat of neutralization ΔH_N. Extensive published material is available on how to determine it with high accuracy by the application of calorimetric or of electrochemical methods.

In solutions with reactants in finite concentration, the heat of the reaction (4.1) depends on the concentration and is larger than ΔH_N^0, that is, larger than the reaction enthalpy in infinitely dilute solution.

4.1.1 CALORIMETRIC DETERMINATION OF THE HEAT OF NEUTRALIZATION

There are two calorimetric methods for the determination of ΔH_N^0 in higher concentration ranges. The first determines the heat of neutralization in a series of increasingly diluted reactants, with direct linear or parabolic extrapolation of ΔH_N^0. The second method is based on extrapolation of ΔH_N values by using the enthalpies of dilution H_v of acid, base, and salt produced. This second method is schematically illustrated in Fig. 4.1.

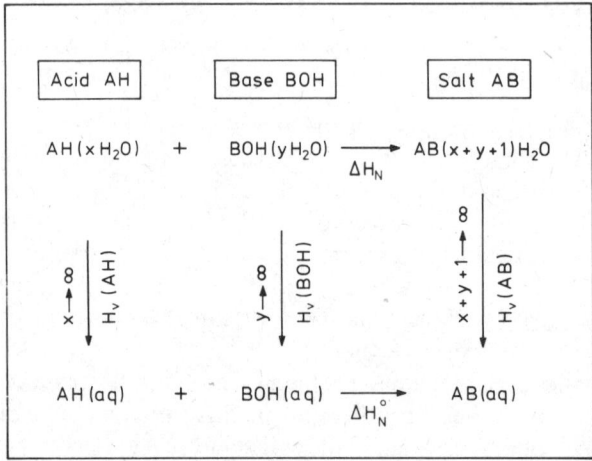

Fig. 4.1. Diagram illustrating the determination of ΔH_N^0.

From Fig. 4.1 we can conclude directly that

$$\Delta H_N^0 = \Delta H_N + H_v(AB) - H_v(AH) - H_v(BOH) \qquad (4.2)$$

Table 4.1 displays the direct parabolic extrapolation of ΔH_N^0 from the measured heat of neutralization ΔH_N in solutions of finite concentration of the reactants, using calorimetric data published by Richards and Rowe.[3] The concentration is given as in Fig. 4.1 as x mol H_2O per mole AH and y mol H_2O per mole BOH.

The mean value of ΔH_N^0 is $-13,693$ cal mol^{-1} at 20°C. With a temperature coefficient[†] $d(\Delta H_N^0)/dt = -50$ cal deg^{-1} we get the value $\Delta H_N^0 = -13,443$ cal mol^{-1} for the heat of neutralization at 25°C.

Bender and Biermann[5] published a critical survey of the experiments and claimed that these were not conducted at sufficiently high dilution to justify direct extrapolation. They determined the value ΔH_N^0 to be $-13,320 \pm 20$ cal deg^{-1} at 25°C from the calorimetric values obtained for two concentrations of the reaction HCl+NaOH ($c=3.239$ m, $\Delta H_N = -14,315$ cal mol^{-1} and $c=3.014$ m, $\Delta H_N = -14,250$ cal mol^{-1}) according

[†]For an exact calculation a squared function of temperature is necessary (see Ref. 4).

THE HEAT EFFECT OF STANDARD REACTIONS

TABLE 4.1. Dependence on Concentration of the Heat of Neutralization and Results of the Direct Parabolic Extrapolation of ΔH_N^0 at $20°C^3$

Reaction	$-\Delta H_N$ cal mol^{-1} $x=y$					$-\Delta H_N^0$ cal mol^{-1}
	25	50	100	200	400	
LiOH + HCl	14 433	14 149	13 993	13 889	13 803	13 685
NaOH + HCl	14 228	14 009	13 895	13 825	13 761	13 660
KOH + HCl	14 569	14 209	14 014	13 905	13 819	13 695
LiOH + HNO$_3$	13 968	13 905	13 863	13 825	13 788	13 715
NaOH + HNO$_3$	14 012	13 892	13 837	13 790	13 756	13 705
KOH + HNO$_3$	14 724	14 325	14 086	13 934	13 834	13 700

to Fig. 4.1. They used the heats of dilution published by Sturtevant[6,7] for HCl and NaOH, and by Robinson for NaCl.[8]

The concentration range of Table 4.1 is extended in Table 4.2; this includes the heat of neutralization observed by these authors for solutions of high reactant concentration.

TABLE 4.2. Dependence on Concentration of the Heat of the Reaction NaOH + HCl at $25°C^5$

HCl/NaOH mol liter^{-1}	$-\Delta H_N$ cal mol^{-1}	HCl/NaOH mol liter^{-1}	$-\Delta H_N$ cal mol^{-1}
3	14 245	10	17 130
4	14 555	11	17 655
5	14 915	12	18 195
6	15 300	13	18 735
7	15 715	14	19 260
8	16 155	15	19 825
9	16 630	16	20 415

The values obtained by Richards and Rowe (Table 4.1) agree well with those of Bender and Biermann (Table 4.2) in the concentration range where they overlap.

To sum up, it can be said that the observed values published by various authors since about 1930 are sufficiently accurate because the most important sources of error had been discovered and were taken into account. Deviations in ΔH_N^0 values are usually due to the extrapolation method used. If the heats of neutralization published by Richards and Rowe and the heats of dilution according to Refs. 6 to 8 are used for 2.222 and 1.111 M solutions corrected to 25°C, values of $-13,323$ and $-13,329$ cal mol^{-1}, respectively, are obtained at 25°C. A new calculation of Pitzer's data (Ref. 4) yielded $-13,353$ cal mol^{-1},[5] and a function published by Rossini[10] gives a value of $-13,320$ cal mol^{-1}. Vanderzee and Swanson[11] obtained $-13,336 \pm 18$ cal mol^{-1} for the reaction $HClO_4 + NaOH$. A recalculation of older values (Ref. 3, 4, 5 and 9) by these authors, using newly determined heats of dilution, gave a mean value $\Delta H_N^0 = -13,340$ cal mol^{-1} at 25°C.

In 1970 Grenthe, Ots, and Ginstrup[12] redetermined the heat of neutralization at various temperatures in the range of 5 to 50°C, using the heats of dilution of the National Bureau of Standards.[13,14] They obtained a value $\Delta H^0 = -55.84 \pm 0.05$ kJ mol^{-1} ($= -13.34 \pm 0.01$ kcal mol^{-1}) at 25°C. The same authors give the following equation for the dependence on temperature of the heat of neutralization ΔH^0 (J mol^{-1}) and T in (K):

$$\Delta H^0 = 4.16578 \cdot 10^5 - 2.74449 \cdot 10^3 \cdot T + 7.031568\, T^2$$
$$- 6.321501 \cdot 10^{-3} \cdot T^3$$

4.1.2 ELECTROCHEMICAL DETERMINATION OF THE HEAT OF NEUTRALIZATION

The electrochemical methods for determination of the heat of neutralization are based on the temperature dependence of the dissociation constant K_w of water

$$\frac{d \ln K_w}{dT} = -\frac{\Delta H_N}{RT^2} \qquad (4.3)$$

The emf method for measuring K_w was developed by Harned and Owen to a high degree of precision.[15] The observed value $\Delta H_N^0 = -13,510$ cal mol^{-1} at 25°C is, with one calorimetric exception,[16] distinctly higher than the calorimetric values discussed in the preceding section. The calorimetric values and the electrochemical values are compared and criticized in Ref. 11.

4.1.3 PROTONIZATION OF TRISHYDROXYMETHYL-AMINOMETHANE AS A NEW STANDARD REACTION

Grenthe, Ots, and Ginstrup[12] measured the heat effect of protonization of trishydroxymethylaminomethane (THAM) at 25°C and its dependence on temperature,

$$\Delta H^0 = -47.44 \pm 0.05 \text{ kJ} \; (= -11.34 \pm 0.01 \text{ kcal mol}^{-1})$$

as a new standard reaction. Further ΔH^0 values are reported by Nelander[18] (-11.36 kcal mol^{-1}) and by Hansen and Lewis[19] (-11.32 kcal mol^{-1}). Trishydroxymethylaminomethane, in the reaction

$$\text{THAM(aq)} + \text{H}^+(\text{aq}) \rightarrow \text{THAM} \cdot \text{H}^+(\text{aq})$$

offers the following advantages as a standard over the neutralization reaction:[17]

1. It can be obtained in a pure form (99.94%).
2. Its reaction with the atmosphere is negligible.
3. It can be easily dried at 100°C and weighed in the air.
4. It has a high equivalent weight (121.137 g).
5. It is available from the National Bureau of Standards as a certified calorimetric standard.
6. It causes no change in ionic strength when titrated with HCl.

The suggested advantage of THAM over NaOH, the possibility of standardizing the titrant by use of the thermometric end point, is according to Hansen and Lewis[19] not a valid criterion for judgment, since the technique is not accurate enough; the NaOH + HClO$_4$ system is for them a better standard because of the possible effects of absorbed CO$_2$.

4.2 TITRATION OF STRONG ACIDS WITH STRONG BASES IN AQUEOUS SOLUTIONS

The titration diagram of the reaction of a strong acid with a strong base is a thermogram of the type in Fig. 2.4. Titration results can be reproduced to 0.2%. On the basis of the thermogram and with the relevant heats of dilution,[6,8] the heat of neutralization[20] emerges as $\Delta H_N^0 = -13.35 \pm 0.08$ kcal mol^{-1}, which agrees well with the calorimetric value of -13.34 kcal mol^{-1}. Hale, Izatt, and Christensen[21] obtained similarly good values with their titration calorimeter. Titrations of HCl and HClO$_4$ with NaOH both resulted in a ΔH_N^0 value of -13.34 ± 0.02 kcal mol^{-1}.

Linde, Rogers, and Hume[22] tested their automatic titration apparatus in a series of NaOH titrations, using various titrand concentrations and solvents (water, 2 N KCl solution, and emulsions). The standard deviation in aqueous solutions was always less than 0.2% with an optimum functioning of the equipment. Becker et al.[23] tested their automatic apparatus using 0.02 N hydrochloric acid at 25°C. They achieved an accuracy of 0.02% and found a heat of neutralization $\Delta H_N \approx \Delta H_N^0 = -13.30 \pm 0.05$ kcal mol^{-1}.

The influence of ionic strength on the titration reaction HCl+NaOH through the addition of various salts (NaCl, KNO$_3$, Na$_2$CO$_3$, Na$_2$SO$_4$, K$_2$SO$_4$, K$_2$CrO$_4$, K$_3$Fe(CN)$_6$, K$_4$Fe(CN)$_6$) is reported by McLean and Penketh.[24] Salts at ionic strengths I of above 0.6 exert a specific influence on the heat of neutralization; this dependence on the concentration of added salts cannot be described as a function of $\sqrt{I}$ as would be expected from the theory of electrolyte solutions.

4.3 TITRATION OF WEAK ACIDS OR WEAK BASES IN AQUEOUS SOLUTIONS

4.3.1 WEAK MONOACIDS AND MONOBASES

The thermometric titration of a weak acid with a base is theoretically complex even in the simplest case, namely, the titration of a weak monoacid with a strong monobase.

The titration of a weak acid AH with a strong base takes place according to the following equations:

$$AH \rightleftharpoons A^- + H^+ \qquad (4.4a)$$

$$H^+ + OH^- \rightleftharpoons H_2O \qquad (4.4b)$$

The equilibrium of the dissociation according to (4.4a) is governed by the dissociation constant k_a

$$k_a = k_d^{(AH)} = \frac{[A^-][H^+]}{[AH]} \qquad (4.5)$$

From the discussion in Section 1.3.3 we can conclude that for every point of the titration

$$[H^+] + \frac{bv_0't}{V} = \frac{K_s}{[H^+]} + \frac{a}{1 + \frac{[H^+]}{k_a}} \qquad (4.6)$$

$K_s/k_b[H^+]$ is put as $\ll 1$ in (1.61), since the base is practically completely dissociated and $bv_0't/V$ is the concentration of the base in the titration vessel. The degree of dissociation α can be derived from (1.64) as

$$\alpha = \frac{k_a}{k_a + [\mathrm{H}^+]} \tag{4.7}$$

It is a function of the volume $v = v_0't$ of titrant B added or of the time t, since the H^+ concentration changes with the addition according to (4.6).

The integral heat effect includes the heats of dissociation and of neutralization of the reactions according to (4.4a and b).

If the size of the dissociation constant is such that dissociation of the acid is complete at the point $bv_0't_\alpha = aV_0$ and all the protons are thus neutralized, we obtain[20]

$$Q_R(t_\alpha) = -bv_0't_\alpha[\Delta H_N + (1-\alpha_0)\Delta H_D] \tag{4.8}$$

It must not be forgotten that only a part of the heat of dissociation is measured because some of the acid is already dissociated when the reaction starts. This can be calculated from (4.6) and (4.7). Using 0.05 N monochloroacetic acid ($k_a = 1.4 \cdot 10^{-3}$) as the titrand, for example, we get $\alpha_0 = 0.15$.

A general description of the part played by the heat of dissociation in the integral heat effect is given in the following equation:

$$Q_R^{(D)}(t) = -aV_0[\alpha(t) - \alpha_0]\Delta H_D \tag{4.9}$$

Equation (4.8) can be derived only if $\alpha(t_\alpha) = 1$. If $\alpha(t_\alpha) < 1$, as for instance in the case of boric acid, for which $\alpha_0 \approx 0$ ($k_a = 5.8 \cdot 10^{-10}$), we obtain instead of (4.8)

$$Q_R(t_\alpha) = -bv_0't_\alpha \alpha(t_\alpha)(\Delta H_N + \Delta H_D) \tag{4.10}$$

and there is no sharp end point. The end point can, however, be extrapolated from the adjoining branches of the curve, since the equilibrium is far enough on the side of the reaction products. The extrapolation of the end point becomes dubious, however, for a titration $\mathrm{Na_2HPO_4 + NaOH}$ ($k_a = 1.2 \cdot 10^{-12}$).

Determination of the heat of dissociation by thermometric titration has usually been based on the following equation:

$$\Delta H_R = \Delta H_N + \Delta H_D \tag{4.11}$$

This approximation results from the limiting case of $\alpha_0 = 0$ and $\alpha(t_\alpha) = 1$; the degree of this approximation must be examined in each individual case. Another limiting case, $\alpha_0 = 1$ and $\alpha(t_\alpha) = 1$, occurs if the titrand is a strong monoacid.

The titration of a weak monoacid of $k_a > 10^{-10}$ in aqueous solution leads to no particular form of thermogram. The thermogram belongs to the type in Fig. 2.4 of a titration with complete titration reaction. The reason for this is easy to see. The enthalpy of dissociation of a monobasic acid is in the range of $|\Delta H_D| < 5$ kcal mol^{-1}. If we take into account the ΔH_N value of -13.4 kcal mol^{-1}, the resulting reaction enthalpy determining the thermogram is $8.5 < \Delta H_R < 13.5$ kcal mol^{-1}.

Consequently, titrations of weak acids are easily possible with an accuracy of 1% without special precautions down to a concentration of 10^{-3} mol liter^{-1}. The end-point indication is sufficiently sharp and is better, especially with very weak acids, than indication with the help of indicators or electrochemical methods, as has been repeatedly demonstrated with the classical example of boric acid (see Fig. 1.1; see also Refs. 25 and 26). End-point detection by thermometric titration can be improved if mannitol is added to the solution of boric acid, as is usually done in the potentiometric method.[26]

Table 4.3 contains a selection of data obtained in thermometric titration of weak monoacids with strong bases.

Avedikian[30] published a report on the titration of acetic acid, propionic acid, and 13 halocarboxylic acids derived from them. Amino acids also have been successfully titrated with NaOH in aqueous solution.[31,32,33]

In the titration of monoacids of $k_a < 10^{-10}$, the shape of the thermogram gradually changes from that of Fig. 2.4 for complete reactions to that of Fig. 2.5 for incomplete reactions according to Fig. 2.8 and depending on the degree at which k_a decreases. Phenols, naphthols, cresols, and so forth, cannot be titrated thermometrically in aqueous solutions with a strong

TABLE 4.3. Results Obtained in Thermometric Titrations of Weak Monoacids With NaOH

Acid	Temperature (°C)	$a \cdot 10^2$ (mol liter^{-1})	k_a	ΔH_R	ΔH_D (kcal mol^{-1})	According to equation	Reference
H$_3$BO$_3$	25	0.1–10	$5.8 \cdot 10^{-10}$	-10.6	$+3.0$	(4.11)	28
	25	5.0		-10.3	$+3.1$	(4.10)	29
CH$_3$COOH	25	0.1–10	$1.8 \cdot 10^{-5}$	-13.5	-0.1	(4.11)	28
CH$_3$(CH$_2$)$_2$COOH	22.8	4.60	$1.5 \cdot 10^{-5}$	-14.05	-0.56	(4.8)	20
CH$_2$ClCOOH	25	0.1–10	$1.6 \cdot 10^{-5}$	-14.4	-1.5	(4.11)	28
	22.3	5.52	$1.4 \cdot 10^{-5}$	-14.47	-1.1	(4.8)	20
CCl$_3$COOH	25	0.1–10	3.10^{-1}	-13.4		(4.11)	28

base. Their sodium salts can, however, be coupled with diazonium salts. This reaction allows the determination of the end point through thermometric titration with an accuracy[27] of better than 1% in the range of 0.05 M to 0.5 M. Equally good results are obtained conductometrically, but the results of potentiometric titrations are not satisfactory.

Titrations of weak bases with a strong acid have also been performed; examples are those of NH_4OH[22,34] and of amines, pyridine, and nicotine,[35] and other organic nitrogen compounds[35,36] with HCl. The titration of a weak acid with a weak base is also successful. The titration of boric acid or acetic acid gives end points for $H_2BO_3NH_4$ and CH_3COONH_4 in spite of the marked hydrolysis of the salts.[37]

Correlations between the thermodynamic data of titration reactions and the structure of the reactants are discussed in Refs. 30, 32, and 38.

4.3.2 MIXTURES OF MONOACIDS AND MONOBASES

The conditions under which the different compounds of a multicomponent system can be determined were discussed in Section 2.3.8. The indication of the end point for each individual component as a break in the curve of the thermogram is possible only in case 2b. The close proximity of ΔH_R values of monoacids makes it highly probable from the very start that a titration of mixtures of acids with a base with the aim of determining their proportion is possible in a limited number of cases only. The condition $\Delta G_1 \neq \Delta G_2$ implies that the dissociation constants must differ sufficiently. If they do not, the break in the curve is ill-defined, since dissociation of the second proton begins before that of the first has come to an end.

A quantitative theory of thermometric neutralization titration of mixtures of acids can be developed on the basis of Section 1.3.3. The degrees of dissociation $\alpha^{(1)}$ and $\alpha^{(2)}$ of two monoacids are defined by (1.64), and the H^+ concentration during the progress of neutralization, by (1.63). The integral heat effect is the sum of the heat effects of the neutralization and dissociation reactions.

A general presentation is not given because the necessary calculations make it imperative to write computer programs for each individual case. This is, however, not very difficult on the basis of the strategy just outlined. In this connection we discuss only some aspects of the titration of a binary mixture of acids, taking as an example the case of a strong and a weak acid. Equation (1.63) defines for the titration with a strong base:

$$[H^+] + b = \frac{K_s}{[H^+]} + a_1 + \frac{a_2}{1 + \frac{[H^+]}{k_{a2}}} \quad (4.12)$$

where a_1 and a_2 are the concentrations of the two titrands and k_{a2} is the dissociation constant of the weak acid. For the strong acid $[H^+]/k_{a1} \ll 1$. The degree of dissociation $\alpha^{(2)}$ is defined by (1.64); $\alpha^{(1)}$ is taken to be $\alpha^{(1)} = 1$.

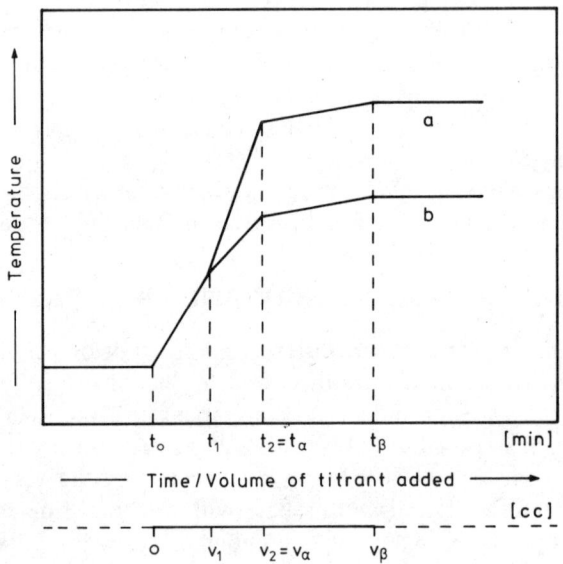

Fig. 4.2. Thermogram of the titration of an equimolar mixture of a strong and a weak acid with sodium hydroxide: (a) ΔH_D of the weak acid <0 (example: HCl—CH$_2$ClCOOH); (b) $\Delta H_D > 0$ (example: HCl—H$_3$BO$_3$). $t_1(v_1)$: equimolar point of the strong acid; $t_\alpha(v_\alpha)$: end point of the titration; and $t_\beta(v_\beta)$: end of the dilution period.

Figure 4.2 illustrates the titration of an equimolar mixture of the acids ($a_1 = a_2$). The strong acid (in this case HCl) suppresses the dissociation of the weak acid ($k_{a1} \gg k_{a2}$). This means that there are two branches during the reaction period if $\Delta H_R^{(2)} - \Delta H_R^{(1)}$ is sufficiently large. Under the conditions stated we get

$$\Delta H_R^{(2)} - \Delta H_R^{(1)} = \Delta H_D^{(2)} \qquad (4.13)$$

Table 4.4 shows how accurately the two end points can be evaluated with the acid mixture.

Boric acid can easily be titrated alongside hydrochloric acid because their heats of reaction differ considerably (see also Ref. 39); even better conditions are encountered in a mixture of boric and sulfuric acid,[26] as shown in Table 4.5.

TITRATION OF WEAK ACIDS OR WEAK BASES IN AQUEOUS SOLUTIONS

TABLE 4.4. Degree of Accuracy of End-Point Indication in Titration of Acid Mixtures with NaOH

Mixture	k_a of the Weak Acid	$\Delta H_R^{(1)} - \Delta H_R^{(2)}$ (kcal mol^{-1})	Accuracy In t_1 %	Accuracy In $t_2 = t_\alpha$ %
HCl–CH$_3$COOH	$1.8 \cdot 10^{-5}$	-0.1	Cannot be recognized	0.2
HCl–CH$_3$(CH$_2$)$_2$COOH	$1.5 \cdot 10^{-5}$	-0.7	± 5	0.2
HCl–CH$_2$ClCOOH	$1.6 \cdot 10^{-5}$	-1.1	± 3	0.2
HCl–H$_3$BO$_3$	$5.8 \cdot 10^{-10}$	$+3.3$	± 1	0.3

TABLE 4.5. Sequential Titration of Sulfuric acid and Boric acid (constant amount of boric acid).[26] Solution titrated: 1.00 ml of 0.102 N H$_3$BO$_3$ + indicated test portion of 0.908 N H$_2$SO$_4$ + H$_2$O to make up to 5.00 ml^a

Test Portion of 0.908 N H$_2$SO$_4$ ml	NaOH Required, meq H$_2$SO$_4$ Titer	NaOH Required, meq H$_3$BO$_3$ Titer
—	—	0.098 (0.102)
0.05	0.039 (0.045)	0.101
0.10	0.089 (0.091)	0.102
0.20	0.178 (0.182)	0.103
0.50	0.461 (0.454)	0.103

^{a}The values in parentheses were calculated from the known titer of the titrand.

The titration of binary mixtures of amino acids with NaOH[31] and of amides, pyridine, and nicotine[35] gives equally good results.

To determine the heats of dissociation of weak acids, it is often advantageous to titrate them mixed with a strong acid because in this case the total heat of dissociation enters into the heat balance of the reaction.

A similar problem arises in the titration of acids in presence of hydrolyzable cations. The heats of reaction of the two titration reactions differ widely as a rule. The titration for determining acid in aqueous hydrofluoric solution of ZrO^{2-} is based on the reactions[40]

$$HF + NaOH \rightarrow NaF + H_2O; \Delta H^0 = -16.27 \text{ kcal mol}^{-1\dagger}$$

$$ZrOF_2 + 2NaOH \rightarrow ZrO(OH)_2 + 2NaF; \Delta H^0 = -4.86 \text{ kcal mol}^{-1}$$

The titration is accurate to 1 to 2%.

Uranium (IV), iron (II), iron (III), and aluminum (III) interfere additively when present in a test solution. No discernible inflection in the titration curve occurs between the neutralization of the free acid and the hydrolysis of these ions. The systems HNO_3–$Th(NO_3)_4$ and H_2SO_4–UO_2SO_4 have also been analyzed successfully.[40] Thermometric titration is often to be preferred for determining the "free-acid content" (for this concept, compare Ref. 41), especially on account of its experimental simplicity.

From titrations of a wide variety of basic nitrogen compounds with 5 N HCl Vaughan and Swithenbank[36] distinguish in their "direct titration method" two categories of thermograms: (1) those with a steep reaction branch, indistinguishable from that of ammonia (Fig. 4.3a); and (2) those with a less steep reaction branch, indistinguishable from that of pyridine (Fig. 4.3b).

Category 1 comprises ammonia, methylamine, diethylamine, trimethylamine, pyrrolidine, piperidine, cyclohexylamine, ethylenediamine, and ethanolamine. Category 2 comprises pyridine, 3-picoline, quinoline, aniline, and o-toluidine. The authors show that it is possible to titrate thermometrically in aqueous solution both aliphatic (category 1), including ammonia, and aromatic (category 2) bases separately in admixture (Fig. 4.3c).

Vaughan and Swithenbank[36] use 5 N NaOH in indirect titration of ammonium chloride and pyridine hydrochloride in the presence of excess acid. After neutralization of the last named, pyridine hydrochloride is first titrated, yielding a steeper reaction branch than ammonium chloride. A mixture of excess acid, pyridine, morpholine, and ethanolamine gives individual end points for each compound.

The direct titration method has been used to determine the ammonia content of several tar works products, some containing pyridine bases that were also determined; the indirect titration method has been preferred for determining pyridine bases in a number of such products that contained ammonia or the free acidity and total pyridine base content of sulfuric acid extracts of base-containing tar oils.[36]

---

†Hydrofluoric acid exists in aqueous solution as HF, HF_2^-, and F^- [W. A. Roth, *Ann.*, **35**, 542 (1939)]. This fact explains the apparent deviation of the ΔH^0 value from ΔH_N^0.

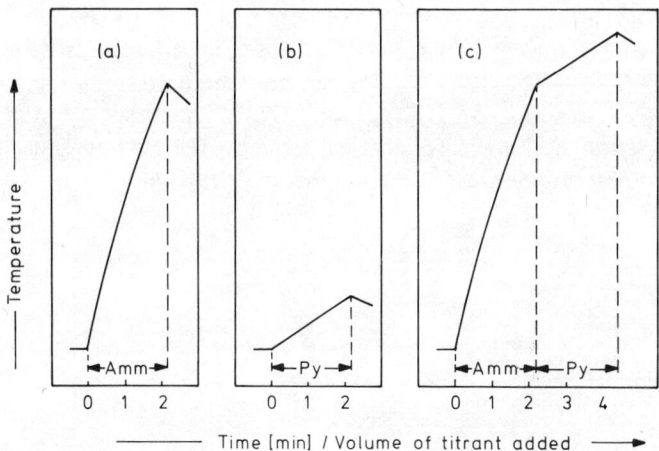

Fig. 4.3. Thermograms of titrations of aqueous solutions of nitrogen bases with 5 N hydrochloric acid[36]: (*a*) ammonia; (*b*) pyridine; (*c*) ammonia + pyridine (equimolar mixture).

4.3.3 POLYVALENT ACIDS

From the analysis of mixtures of monoacids it can be concluded that polyvalent acids will not necessarily produce the same number of breaks in the thermogram, as they possess acidic H atoms.

The titration of sulfuric acid H_2SO_4 with NaOH produces an end point accurate to 0.2% for Na_2SO_4, but the break in the curve for $NaHSO_4$ cannot be located with sufficient precision (see Ref. 26), although something like an end point has been observed.[2] In an extremely careful investigation, Christensen et al.[42] determined from a thermogram the heat of dissociation and the pK value of the reaction $HSO_4^- \rightleftharpoons H^+ + SO_4^{2-}$ as $\Delta H^0 = -5.64 \pm 0.08$ kcal mol^{-1} and p$K = 1.91 \pm 0.01$.

The types of titration diagram for the neutralization of polyvalent acids will be discussed using two examples: citric acid–sodium hydroxide and phosphoric acid–sodium hydroxide.

In a titration under the experimental conditions of Fig. 4.4, citric acid produces no breaks in the ΔT-t diagram except the end point of the entire reaction. An evaluation of the titration diagram shows that this is due to the combined effects of the heats of dissociation and the dissociation constants, which differ only slightly for the successive steps.

The first step in the evaluation of the thermogram in Fig. 4.4 is the determination of the titer of citric acid with the help of the end point ($t_3 = t_\alpha$). Then two more points are chosen, in the present case $t_1 = t_\alpha/3$ and $t_2 = 2t_\alpha/3$. For the points (t_0, T_0), (t_1, T_1), (t_2, T_2), and (t_α, T_α) the hypothetical pre- and afterperiods (00'), (11'), (22'), and ($\alpha\alpha'$) are constructed according to the rules given in Section 2.3.4. The differences in temperature $T_1^* - T_0^*$, $T_2^* - T_1^*$, and $T_\alpha^* - T_2^*$ then give the corresponding values of the enthalpy of reaction ΔH_R^I, ΔH_R^{II}, and ΔH_R^{III}. The dissociation constants are presumed to be known for this experiment. If necessary, they can be determined potentiometrically in a parallel experiment.[†]

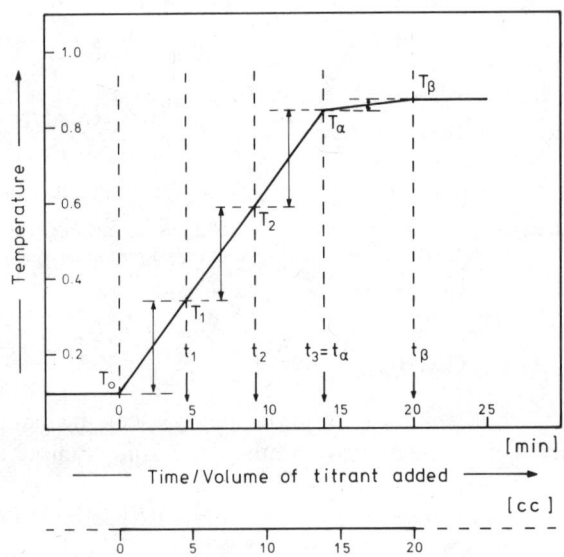

Fig. 4.4. Thermogram of the titration of citric acid with sodium hydroxide ($V_0 = 200$ ml, $b = 0.930$ M).[20]

The degrees of dissociation $\alpha(t)$, $\beta(t)$, and $\gamma(t)$ can be calculated for every point of the titration curve from the dissociation constants and the known concentration of citric acid [see Table (4.6)].

The general system of equations for a tribasic acid with complete titration is

$$\Delta H_R^I = (\alpha_1 - \alpha_0)\Delta H_D^{(1)} + (\beta_1 - \beta_0)\Delta H_D^{(2)} + (\gamma_1 - \gamma_0)\Delta H_D^{(3)} + \Delta H_N$$

$$\Delta H_R^{II} = (\alpha_2 - \alpha_1)\Delta H_D^{(1)} + (\beta_2 - \beta_1)\Delta H_D^{(2)} + (\gamma_2 - \gamma_1)\Delta H_D^{(3)} + \Delta H_N$$

$$\Delta H_R^{III} = (1 - \alpha_2)\Delta H_D^{(1)} + (1 - \beta_2)\Delta H_D^{(2)} + (1 - \gamma_2)\Delta H_D^{(3)} + \Delta H_N$$

[†]They were calculated for the temperature of the experiment with a three term pK equation and the values given in Ref. 43 (Appendix 12.1).

TITRATION OF WEAK ACIDS OR WEAK BASES IN AQUEOUS SOLUTIONS

TABLE 4.6. Dissociation of Citric Acid

t	0		$t_1 = t_\alpha/3$			$t_2 = 2t_\alpha/3$			$t_3 = t_\alpha$	
Ions	AH_2^-		AH_2^-	AH^{2-}		AH_2^-	AH^{2-}	A^{3-}	AH^{2-}	A^{3-}
Degree of Dissociation	α_0	$\beta_0 = 0$	α_1	β_1	$\gamma_1 = 0$	$\alpha_2 = 1$	β_2	γ_2	$\beta_3 = 1$	$\gamma_3 = 1$

The degree of dissociation of citric acid shown in Table 4.6 and the data of Fig. 4.4 must be inserted in this system of equations, and we then obtain

$$\Delta H_D^{(1)} = 1.20 \text{ kcal mol}^{-1} \quad \Delta H_D^{(2)} = 0.70 \text{ kcal mol}^{-1}$$

$$\Delta H_D^{(3)} = -0.78 \text{ kcal mol}^{-1}$$

These results can be compared with the values (25°C) published by Bates and Pinching:[44]

$$\Delta H_D^{(1)} = 0.997 \text{ kcal mol}^{-1} \quad \Delta H_D^{(2)} = 0.583 \text{ kcal mol}^{-1}$$

$$\Delta H_D^{(3)} = -0.803 \text{ kcal mol}^{-1}$$

When titrated in aqueous solution with sodium hydroxide, phthalic, oxalic, malonic, and succinic acids produce thermograms showing two breaks for the successive neutralization of the acid groups; glutaric, adipic, maleic, fumaric, and tartaric acids show no intermediate arrests.[45]

Other investigations of divalent carboxylic acids are found in Jordan and Dumbaugh:[46] oxalic acid, malonic acid, succinic acid, glutaric acid, adipic acid, pimelic acid, and azelaic acid.

A second typical thermogram displaying different characteristics is that produced by the titration of phosphoric acid (see Fig. 4.5). If the dissociation constants are known,[43,47] it is not difficult to recognize the first and second protons of phosphoric acid. The breaks seen in Fig. 4.5 imply that the heats of dissociation differ sufficiently. It has already been shown that the reaction of phosphoric acid is not complete in (t_α, T_α), and not even in (t_β, T_β) when a 100% excess of base is present. The titer of the phosphoric acid must hence be calculated from one of the other two breaks in the titration curve, preferably from (t_2, T_2). It must be noted that the diagram can have no period of dilution so that another unknown quantity, the heat of dilution H_v of the sodium hydroxide solution, is added to the three unknowns $\Delta H_D^{(1)}$, $\Delta H_D^{(2)}$, and $\Delta H_D^{(3)}$ of a trivalent acid with complete

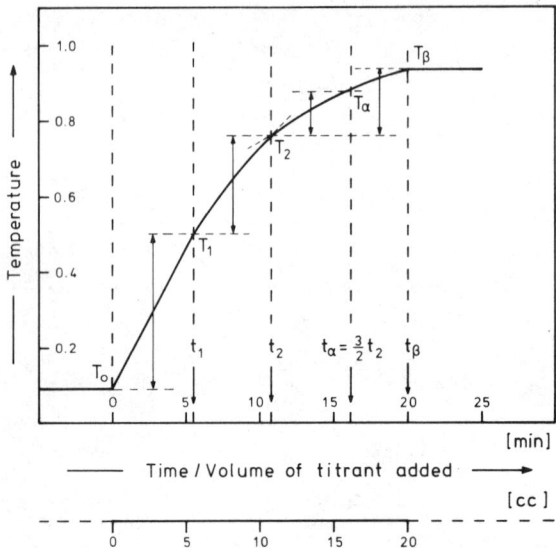

Fig. 4.5. Thermogram of the titration of phosphoric acid with sodium hydroxide ($V_0 = 200$ ml, $b = 0.930$ M).[20]

titration. It is, therefore, necessary to have a fourth equation:

$$\Delta H_{R+v}^{I} = \Delta H_N + H_v + (1 - \alpha_0)\Delta H_D^{(1)}$$

$$\Delta H_{R+v}^{II} = \Delta H_N + H_v + \Delta H_D^{(2)}$$

$$\Delta H_{R+v}^{III} = \gamma_3 \Delta H_N + H_v + \gamma_3 \Delta H_D^{(3)}$$

$$\Delta H_{R+v}^{IV} = \frac{t_\alpha - t_2}{t_\beta - t_2}\gamma_\beta \Delta H_N + H_v + \frac{t_\alpha - t_2}{t_\beta - t_2}\gamma_\beta \Delta H_D^{(3)}$$

The equations are solved with the help of the ΔH_{R+v} values derived from Fig. 4.5 (the index $R+v$ indicates that the reaction heat and the heat of dilution cannot be separated):

$$\Delta H_D^{(1)} = -1.65 \text{ kcal mol}^{-1} \qquad \Delta H_D^{(2)} = 0.78 \text{ kcal mol}^{-1}$$

$$\Delta H_D^{(3)} = 3.12 \text{ kcal mol}^{-1}$$

Published values (25°C) are

$$\Delta H_D^{(1)} = -1.75 \text{ kcal mol}^{-1} \text{ (Ref. 48)} \qquad \Delta H_D^{(2)} = 0.82 \text{ kcal mol}^{-1} \text{ (Ref. 49)}$$

$$\Delta H_D^{(3)} = 3.5 \pm 0.5 \text{ kcal mol}^{-1} \text{ (Ref. 4)}$$

Christensen et al.[42] also state on the basis of their highly accurate studies that the end point of the reaction $HPO_4^{2-} + OH^- \rightleftharpoons PO_4^{3-} + H_2O$ cannot be determined precisely. Dutoit and Grobet,[2] on the other hand, did find it in the thermogram of the neutralization of phosphoric acid, together with the better-defined end points of NaH_2PO_3 and Na_2HPO_3. Linde et al.[22] also located an end point, but it is displaced considerably toward the excess of NaOH and is far less accurately determined than the other two end points.

The titration of arsenic acid H_3AsO_4 with NaOH produces a thermogram with three recognizable breaks.[50] Arsenious acid has also been titrated,[37,51] as well as a number of other phosphoric acids and mixtures of these.

Pâris and Tardy[52] compared the potentiometric, conductometric, and thermometric titrations of a mixture of hypophosphorous acid H_3PO_2 (pK = 1.0), phosphorous acid H_3PO_3 (pK_1 = 1.4, pK_2 = 6.7), and phosphoric acid H_3PO_4 (pK_1 = 2.16, pK_2 = 7.16, pK_3 = 12.4); they state that thermometric titration is the most advantageous method.

Pâris and Robert[53,54] used the thermometric method for the titration of phosphoric acid H_3PO_4, diphosphoric acid $H_4P_2O_7$ (pK_1 = 1, pK_2 = 2.3, pK_3 = 6.7, pK_4 = 9.4), and metaphosphoric acid $(HPO_3)_n$ (pK = 2.0). Metaphosphoric acid reacts as a monobasic, diphosphoric acid as a dibasic, and phosphoric acid as a tribasic acid. They titrated each acid individually, all mixtures of two acids, and finally the mixture of all three acids. This mixture shows three breaks on the curve: (1) the neutralization of the strong acids [pK < 3 for $(NaPO_3)_n$, $Na_2H_2P_2O_7$, and NaH_2PO_4]; (2) the neutralization of the medium acids (3 < pK < 9 for $Na_4P_2O_7$ and NaH_2PO_4); (3) the neutralization of the weak acid (pK = 12.4 for Na_3PO_4]. The information obtained from the end points suffices for an analysis of the mixture. The titration of a mixture of the salts of these acids is discussed in Ref. 55.

Other examples quoted for possible titrations of mixtures of acids[53] are those of hypophosphorous acid H_3PO_2, diphosphoric acid $H_4P_2O_7$, and phosphoric acid H_3PO_4; and of hydrochloric acid, acetic acid, and arsenic acid H_3AsO_4. A titration of phosphoric acid and arsenic acid, on the other hand, is not possible.

Dibasic compounds have been titrated with hydrochloric acid by Popper, Roman, and Marcu.[56] The neutralization curves of the titration of o-, m-, and p-phenylenediamine show both dissociation steps. A special characteristic of this and an earlier investigation[57] of the titration of two weakly acidic sulfur compounds is the use of oblique-angled coordinates for the titration diagram to eliminate the heat of dilution.

Van Dalen and Ward[58] report a rapid and exact thermometric titration determination of hydroxide and alumina in Bayer process solutions, using 1.3 N HCl; the other constituents in Bayer liquors do not interfere.

4.3.4 DISPLACEMENT OF A FUNCTION

Titrations of salts of weak acids with strong acids were carried out for Na_3PO_4, Na_2HPO_4, and NaH_2PO_4 with HNO_3;[2] and for Na_2CO_3, $NaHCO_3$, $Na_2B_4O_7$, $Na_2S_2O_3$, and Na_3PO_4 with HCl.[34] As an example, Fig. 4.6 shows the thermogram of the titration of $Na_2B_4O_7 \cdot 10H_2O$ with HCl.[29]

An example of a complex neutralization is reported by Vaughan.[59] The titration of an equimolar mixture of sodium hydroxide, sodium carbonate, and sodium sulfite with hydrochloric acid yields a thermogram with five inflections in the titration period: sodium hydroxide, the first and second ionization of the sodium carbonate, and the first and second ionization of the sodium sulfite.

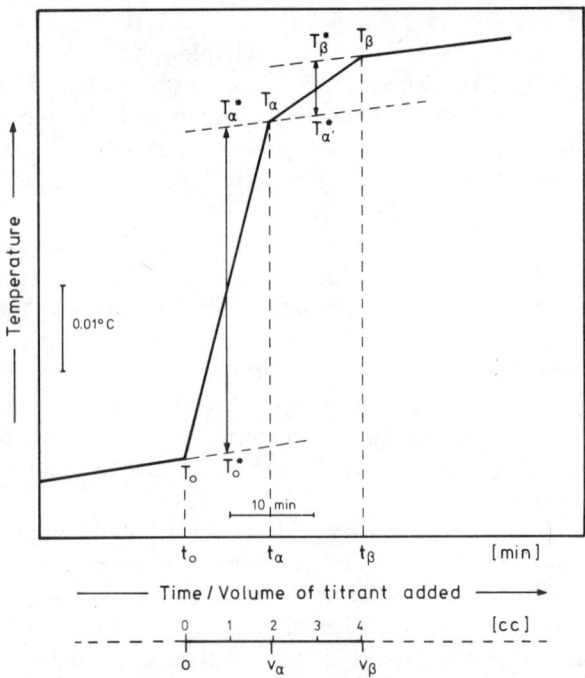

Fig. 4.6. Thermogram of the titration of $Na_2B_4O_7$ with HCl ($V_0 = 185$ ml, $a = 0.005\ M$, $b = 1\ M$).[29]

The indirect titration method of Vaughan and Swithenbank for the titration of basic nitrogen compounds with NaOH is another example (Section 4.3.2). Sen and Wu[33] also titrated basic amino acids as their hydrochlorides with NaOH. De Leo and Stern[60] used thermometric endpoint indication for displacement reactions of weak-base salts with NaOH in the analysis of pharmaceuticals, such as chlorpheniramine maleate, chlorpromazine hydrochloride, and hydrochlorothiazide. Beezer and Slawinsky titrated codeine phosphate with alkali.[61]

4.3.5 FORMATION AND DECOMPOSITION OF POLYANIONS

Kiba and Takeuchi[62] used perchloric acid for thermometric titration of solutions of sodium molybdate, sodium tungstate, sodium orthovanadate, ammonium metavanadate, and potassium chromate in a twin-cell titrator (see Chapter 9) and compared thermometric and potentiometric titration curves. In the course of titration the formation of polyanions is observed. Methods have been developed for the determination of tungsten, vanadium, and chromium.

Spitsin and Kosmodem'yanskaya[63-66] studied the decomposition of polytungstates by titration with sodium hydroxide solution. Tungstophosphoric acid, for example, is neutralized by the slow addition of three equivalents of NaOH. Rapid addition of the third equivalent or reaction with an excess of NaOH causes decomposition of the heteropolyanion. The reactions taking place are[63]

$$H_3(PW_{12}O_{40}) + 3OH^- \rightleftharpoons (PW_{12}O_{40})^{3-} + 3H_2O$$

$$4(PW_{12}O_{40})^{3-} + 6OH^- \rightleftharpoons 4(PW_9O_{31})^{3-} + 3(W_4O_{13})^{2-} + 3H_2O$$

REFERENCES

1. J. M. Bell and C. F. Cowell, *J. Am. Chem. Soc.*, **35**, 49 (1913).
2. P. Dutoit and E. Grobet, *J. Chim. Phys.*, **19**, 324 (1922).
3. T. W. Richards and A. W. Rowe, *J. Am. Chem. Soc.*, **44**, 684 (1922).
4. K. S. Pitzer, *J. Am. Chem. Soc.*, **59**, 2365 (1937).
5. P. Bender and W. J. Biermann, *J. Am. Chem. Soc.*, **74**, 322 (1952).
6. J. M. Sturtevant, *J. Am. Chem. Soc.*, **62**, 3265 (1940).
7. J. M. Sturtevant, *J. Am. Chem. Soc.*, **62**, 2276 (1940).
8. A. L. Robinson, *J. Am. Chem. Soc.*, **54**, 1311 (1932).
9. L. J. Gillespie, R. H. Lambert, and J. A. Gibson, Jr., *J. Am. Chem. Soc.*, **52**, 3806 (1930).
10. F. D. Rossini, *J. Res. Natl. Bur. Stand.*, **6**, 791 (1931).

11. C. E. Vanderzee and J. A. Swanson, *J. Phys. Chem.*, **67**, 2608 (1963).
12. I. Grenthe, H. Ots, and O. Ginstrup, *Acta Chem. Scand.*, **24**, 1067 (1970).
13. D. D. Wagman, S. M. Bailey, and R. H. Schumm, "Selected Values of Chemical Thermodynamic Properties," *Natl. Bur. Stand. Tech. Note 270*, Jan. 3, 1968.
14. F. D. Rossini, D. D. Wagman, W. H. Evans, S. Levine, and I. Jaffe, "Selected Values of Chemical Thermodynamic Properties," *Natl. Bur. Stand. Circ. 500*, Washington, D.C., 1952.
15. H. S. Harned and B. B. Owen, *Chem. Rev.*, **25**, 31 (1939).
16. H. M. Papee, W. J. Canady, and K. J. Laidler, *Can. J. Chem.*, **34**, 1677 (1956).
17. E. W. Wilson, Jr. and D. F. Smith, *Anal. Chem.*, **41**, 1903 (1969).
18. L. Nelander, *Acta Chem. Scand.*, **18**, 973 (1964).
19. L. D. Hansen and E. A. Lewis, *J. Chem. Thermodyn.*, **3**, 35 (1971).
20. J. Barthel, F. Becker, and N. G. Schmahl, *Z. Phys. Chem.* (New Series), **29**, 58 (1961).
21. I. D. Hale, R. M. Izatt, and J. J. Christensen, *Proc. Chem. Soc.*, **1963**, 240.
22. H. W. Linde, L. B. Rogers, and D. N. Hume, *Anal. Chem.*, **25**, 404 (1953).
23. F. Becker, J. Barthel, N. G. Schmahl, G. Lange, and H. M. Lüschow, *Z. Phys. Chem.* (New Series), **37**, 33 (1963).
24. W. R. McLean and G. E. Penketh, *Talanta*, **15**, 1185 (1968).
25. F. Pechar, *Chem. Listy*, **59**, 1073 (1965).
26. F. J. Miller and P. F. Thomason, *Talanta*, **2**, 109 (1959).
27. R. A. Pâris and J. Vial, *Chim. Anal.*, **1952**, 223.
28. J. Jordan and WM. H. Dumbaugh, Jr., *Anal. Chem.*, **31**, 210 (1959).
29. J. Barthel and K. Wachter-Lenz, unpublished.
30. L. Avedikian, *Bull. Soc. Chim. Fr.*, **1966**, 2570.
31. K. K. Chatterji and A. K. Ghosh, *J. Indian Chem. Soc.*, **34**, 407 (1957).
32. L. Avedikian, *Bull. Soc. Chim. Fr.*, **1967**, 254.
33. B. Sen and W. C. Wu, *Anal. Chim. Acta*, **46**, 37 (1969).
34. P. T. Priestley, *Analyst*, **88**, 194 (1963).
35. R. D. Daftary and B. C. Haldar, *Anal. Chim. Acta*, **25**, 538 (1961).
36. G. A. Vaughan and J. J. Swithenbank, *Analyst*, **92**, 364 (1967).
37. P. Mondain-Monval and R. Pâris, *C. R. Acad. Sci. (Paris)*, **207**, 338 (1938).
38. J. Jordan, in I. M. Kolthoff and P. J. Elving, *Treatise on Analytical Chemistry*, Part I, Vol. 8, Interscience, New York, 1968, p. 5175.
39. P. Papoff and P. G. Zambonin, *Ric. Sci.*, **35**, 93 (1965).
40. F. J. Miller and P. F. Thomason, *Anal. Chem.*, **31**, 1498 (1959).
41. G. L. Booman, M. C. Elliott, R. B. Kimball, F. O. Cartan, and J. E. Rein, *Anal. Chem.*, **30**, 284 (1958).
42. J. J. Christensen, R. M. Izatt, L.D. Hansen, and J. A. Partridge, *J. Phys. Chem.*, **70**, 2003 (1966).
43. R. A. Robinson and R. H. Stokes, *Electrolyte Solutions*, Butterworths, London, 1970.
44. R. G. Bates and G. D. Pinching, *J. Am. Chem. Soc.*, **71**, 1274 (1949).
45. R. J. N. Harries, *Talanta*, **15**, 1345 (1968).
46. J. Jordan and W. H. Dumbaugh, Jr., *Bull. Chem. Thermodyn. IUPAC*, No. 2, 1959.

47. C. Hodgman, *Handbook of Chemistry and Physics*, 33rd ed., Chemical Rubber Co., Cleveland, Ohio, 1951.
48. L. F. Nims, *J. Am. Chem. Soc.*, **56**, 1110 (1934).
49. H. D. Everett and W. F. K. Wynne-Jones, *Trans. Faraday Soc.*, **35**, 1380 (1939).
50. P. Mondain-Monval and R. Pâris, *Bull. Soc. Chim. Fr.*, **5**, 1641 (1938).
51. C. Mayr and J. Fisch, *Z. Anal. Chem.*, **76**, 418 (1929).
52. R. Pâris and P. Tardy, *C. R. Acad. Sci. (Paris)*, **223**, 1001 (1946).
53. R. Pâris and J. Robert, *C. R. Acad. Sci. (Paris)*, **223**, 1135 (1946).
54. R. Pâris and J. Robert, *Bull. Soc. Chim. Fr.*, **10**, 224 (1943).
55. R. Pâris and P. Tardy, *Bull. Soc. Chim. Fr.*, **12**, 16 (1945).
56. E. Popper, L. Roman, and P. Marcu, *Talanta*, **12**, 249 (1965).
57. E. Popper, L. Roman, and P. Marcu, *Talanta*, **11**, 515 (1964).
58. E. van Dalen and L. G. Ward, *Anal. Chem.*, **45**, 2248 (1973).
59. G. A. Vaughan, *Thermometric and Enthalpimetric Titrimetry*, Van Nostrand Reinhold, New York, 1973.
60. A. B. De Leo and M. Stern, *J. Pharm. Sci.*, **55**, 173 (1966).
61. A. E. Beezer and A. K. Slawinski, *Talanta*, **18**, 837 (1971).
62. N. Kiba and T. Takeuchi, *Talanta*, **20**, 875 (1973).
63. V. I. Spitsyn and G. V. Kosmodem'yanskaya, *Zh. Neorg. Khim.*, **10**, 353 (1965).
64. V. I. Spitsyn and G. V. Kosmodem'yanskaya, *Zh. Neorg. Khim.*, **11**, 1397 (1966).
65. V. I. Spitsyn and G. V. Kosmodem'yanskaya, *Zh. Neorg. Khim.*, **13**, 2213 (1968).
66. V. I. Spitsyn and G. V. Kosmodem'yanskaya, *Omagiu Raluca Ripan*, **1966**, 581.

CHAPTER

5

TITRATIONS IN NONAQUEOUS SOLVENTS

Several factors determine the choice of a nonaqueous solvent for a titration. Water cannot be used in a number of titrations because it reacts with one of the reactants; there is also the question of solubility in nonaqueous solvents. In addition to these more general considerations, specific qualities of a solvent may favor its use (see Section 5.5).

Naturally, most titrations in nonaqueous solvents are those of Lewis or Brönsted acids or bases. Depending on the basicity of the solvent used, the leveling influence of water, which is highly basic, can be eliminated when mixtures of acids of different basicity are titrated. Thus the titration of a mixture of H_2SO_4 and HCl with lithium acetate in acetic acid produces the following end points as neutralization proceeds: first proton of the sulfuric acid, hydrochloric acid, second proton of sulfuric acid.[1] Diprotic acids, such as maleic acid, which yield thermograms without indication for the neutralization of the first proton when titrated in aqueous solution, show breaks for the successive neutralization of the two protons when titrated in methanol–benzene solutions.[2]

Up to now only few thermometric titrations have been carried out in nonaqueous solvents. A recent survey of titration in nonaqueous solvents[3] mentions 254 titrations, of which only 10 have been with the thermometric method. In thermometric titration no laborious search for a suitable indicator, no preliminary research on the system of electrodes, and so forth are necessary. As the specific heat of most nonaqueous solvents is smaller than that of water, $(\Delta T)_{EP}$ is, as a rule, larger. Considerable heats of dilution and high heats of solution may result, however. This implies a more elaborate method of evaluation of the thermodynamic data of the reaction from the thermograms. Data that are available in tables for aqueous solutions must first be obtained for nonaqueous solutions in separate experiments.

5.1 DETERMINATION OF THE PURITY OF SOLVENTS

A fundamental problem in using nonaqueous solvents is the frequent need for careful drying. The method used by Karl Fischer[4] to determine the water content has been worked out as a thermometric method by Wasilewski and Miller.[5] The authors made use of the markedly exothermic heat of reaction of -16.1 kcal mol^{-1} H$_2$O in a three-stage DIE procedure. Other thermometric methods for the determination of the water content have been proposed by Erdey and Maric,[6] Spink and Spink,[7] and by Greathouse et al.[8] The last named is based on the considerable heat effect of the hydrolysis reaction of acetic anhydride. This reaction is slow, but is catalyzed by strong acids.[9] Water (and acetic anhydride) can, therefore, be determined by thermometric titration in acetic acid in the presence of perchloric acid as a catalyst.

Bark and Bate[10] determine carboxylic acid anhydrides in the presence of the parent acids by addition of the anhydride to a methanolic solution of morpholine and titration of the unreacted morpholine with a methanolic solution of hydrochloric acid.

Acetic acid in acetic anhydride can be determined by titration with triethylamine.[11]

Rogers[12] has published a DIE method for the determination of olefins in hydrocarbons; Crompton and Cope,[13] for the same solvents, have developed a continuous-flow method for determining alcohols, water, and other compounds that react with triethylaluminum.

5.2 LEWIS ACIDS AND LEWIS BASES

Thermograms of thermometric titrations in nonaqueous solution are of the same fundamental types as those in aqueous solution. It must be remembered that even traces of water in the solvent may lead to a noticeable deformation of the titration curve compared with that of a titration in an anhydrous solvent. Figure 5.1 shows the diagram of a titration of SnCl$_4$ with dioxane in benzene as solvent.[14] Anhydrous SnCl$_4$ in benzene is the titrand and is titrated with 1 M dioxane in benzene. The titration is possible with an accuracy of 1%.

This simple method for a solid Lewis acid was first used by Trambouze,[15] who titrated AlCl$_3$ with dioxane and ethyl acetate in benzene. Later, Trambouze et al. used the titration with dioxane or ethyl acetate as a method for determining the Lewis acidity of mixed alumina–silica gels.[16] The Lewis acid AlBr$_3$ was titrated with various ethers in benzene solution by Romm et al.[17]

Zenchelsky et al.[18,19] investigated the titration of SnCl$_4$ in aprotic solvents. As a titrand, SnCl$_4$, in samples from 3 to 90 μmol per 100 ml of solvent (benzene, carbon tetrachloride, nitrobenzene, or chloroform), can be titrated with dioxane in benzene or carbon tetrachloride; the standard deviation of the results is 1%, using nitrobenzene 5%. Chloroform is not suitable for this titration because of its small heat effect. The reaction heat

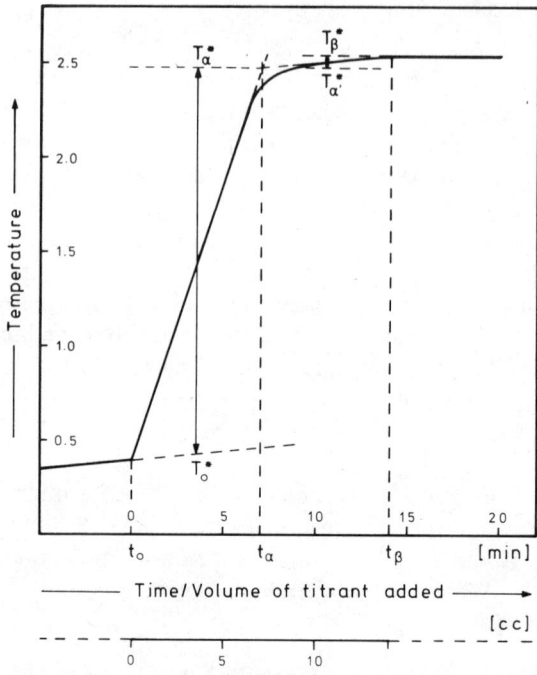

Fig. 5.1. Thermogram of the titration of $SnCl_4$ with dioxane in benzene as solvent[14] ($V_0 = 200$ ml, $b = 1.166$ M).

ΔH_R of the titration reaction[18]

$$SnCl_4(\text{solution}) + C_4H_8O_2(\text{solution}) \rightarrow \text{Adduct} + \text{Solvent}$$

was determined by using calorimetric data for the individual steps of the reaction. Subsequent studies were made of the reaction of $SnCl_4$ with tetrahydrofuran and pyridine[20] and with tetrahydrofuran, tetrahydropyran, 2-methyl- and 2,5-dimethyltetrahydrofuran, and 4-methyltetrahydropyran[21] in benzene, which lead to the formation of addition compounds of the type $SnCl_4 \cdot 2X$. The 1:1 complex with dioxane, the formation of which was utilized in the analyses, is converted into a 1:2 compound in a subsequent reaction. In Chapter 8 the aspects of entropy titration in these publications are discussed. Further, titrations of $SnCl_4$ with diethoxyethane and dibutylthioethane in benzene yield end points that can be analytically evaluated.[22] Further reactions of tin(IV) chloride and/or organotin(IV) chlorides with bases of this type [e.g., $RO(CH_2)_nOR$, $RS(CH_2)_nSR$, R_2S] and bases such as $R'_nSn(OR)_{4-n}$ generally yield 1:1 adducts, which can be used in the quantitative determination of these Lewis acids.[23,24]

The construction of the end point is possible by extrapolation for the reaction of I_2 with dibutyl sulfide in octane. The thermodynamic data of this reaction and other reactions that cannot be evaluated for analytical purposes have been determined.[22] In a more recent work Nelander[25] gives heats of formation and stability constants in ethylene dichloride solution for complexes between iodine and methyl-, ethyl-, propyl-, isopropyl-, butyl-, and *tert*-butyl disulfides, and 1,2-dithiane. N,N-Dimethylacetamide (DMA) reacts with iodine to form DMA–I_2. Enthalpies and equilibrium constants of this reaction in methylene chloride and benzene are reported by Drago et al.[26]

5.3 ORGANOMETALLIC COMPOUNDS

In a comprehensive study Hoffmann and Tornau[27] showed the value of the thermometric method when applied to titrations of organoaluminum compounds with amines and ethers in cyclohexane. The titration curve of pure aluminum triethyl with triethylamine has a sharp break at the equivalence point. The thermogram of the same titration using commercial aluminum triethyl (hydride content 8.1%) shows two sections in the reaction period, making it possible to determine AlR_3 as well as R_2AlH. If only trialkyl is to be determined, this can be done even better by titrating with an ether. Mixtures of AlR_3 and R_2AlCl cannot be evaluated separately. Another type of reaction for the analysis of organoaluminum compounds is the stepwise substitution of alkyl groups by alkoxy groups in a titration with a primary alcohol. Titration with a tertiary alcohol produces only the first equivalence point, but this is particularly sharp. Everson and Ramirez[28] systematically investigated the reactions of alkylaluminum compounds with amines, oxygen-containing compounds, and chelating agents. Triethylamine and also isoquinoline and 2,2'-bipyridine permit the simultaneous determination of R_2AlH and R_3Al. Where ethers can be used as specific reactants for the determination of R_3Al, a subsequent titration with a ketone permits the simultaneous evaluation of the R_3Al and R_2AlH contents of the sample. Titration with oxine yields complex thermograms, the stages of which are discussed by the authors mentioned.

Diethylzinc in hydrocarbons as titrand can be titrated thermometrically with *o*-phenanthroline, even in the presence of its oxidation and hydrolysis products. With 2,2'-bipyridyl the titration is similar but less satisfactory. The compound Et_2Zn and its oxidation products, $EtZnOEt$ and $Zn(OEt)_2$, are titrated quantitatively with 8-quinolinol, whereas the titration of $EtZnOH$, a product of hydrolysis, is incomplete.[29]

Butyl lithium reacts with butanol in hydrocarbon solvents according to the equation $RLi + BuOH \rightarrow LiOBu + RH$. The reaction is complete and has a heat of reaction[30] of about 50 kcal mol^{-1}. This thermometric titration is very accurate. The main impurity, LiOBu, does not interfere because it is the final product of the reaction.

Parker and Vlismas[31] used thermometric titration to determine Grignard reagent activity. The result agreed excellently with those from the acidimetric titration method.

5.4 BRÖNSTED ACIDS AND BRÖNSTED BASES

The properties of glacial acetic acid as an amphiprotic solvent have been demonstrated by the investigations of Kolthoff and Bruckenstein[32-34] and Huber.[35] Keily and Hume[36] studied thermometrically the neutralization reactions of Brönsted bases with perchloric acid in this solvent.

The base B forms ions in its reaction with the solvent according to

$$B + CH_3COOH \rightleftharpoons \underset{\text{ionization}}{BH^+ \cdot CH_3COO^-} \rightleftharpoons \underset{\text{dissociation}}{BH^+ + CH_3COO^-} \quad (5.1)$$

This is a two-step reaction via an ion pair $BH^+ \cdot CH_3COO^-$. The respective equilibrium constants are

$$K^{(B)} = \frac{[BH^+][CH_3COO^-]}{[B] + [BH^+ \cdot CH_3COO^-]} \quad (5.2)$$

$$K_I^{(B)} = \frac{[BH^+ \cdot CH_3COO^-]}{[B]} \quad (5.3)$$

$$K_D^{(B)} = \frac{[BH^+][CH_3COO^-]}{[BH^+ \cdot CH_3COO^-]} \quad (5.4)$$

The overall dissociation constant K is, therefore, made up of the constant of ionization K_I and the dissociation constant K_D

$$K = \frac{K_I K_D}{1 + K_I} \quad (5.5)$$

The essential difference between water and glacial acetic acid as solvents is that ionization reactions do not play any part in water. This may lead to a change in the basic strength order in glacial acetic acid compared to water. For a weak base in glacial acetic acid $K = K_I \cdot K_D$; for a strong base $K = K_D$.

Similar conditions apply to perchloric acid, which reacts according to

$$HClO_4 + CH_3COOH \rightleftharpoons \underset{\text{ionization}}{CH_3 \cdot COOH_2^+ \cdot ClO_4^-} \rightleftharpoons \underset{\text{dissociation}}{CH_3COOH_2^+ + ClO_4^-}$$

$$(5.6)$$

with corresponding equilibrium constants. The solvent is regarded as anhydrous. In the presence of water, a stronger base than acetic acid, only H_3O^+ is formed so that two neutralization reactions take place simultaneously,

$$CH_3COOH_2^+ + B \rightleftharpoons BH^+ + CH_3COOH \qquad (5.7)$$

$$H_3O^+ + B \rightleftharpoons BH^+ + H_2O \qquad (5.8)$$

Keily and Hume determined the reaction enthalpies of titrated acetates (Pb^{2+}, K^+, Ba^{2+}, Cd^{2+}, Li^+, Na^+, Mg^{2+}, Hg^{2+}) with perchloric acid in glacial acetic acid as solvent, using the initial slope method. They compared two groups of results, those from titration with anhydrous perchloric acid according to (5.7), and those using hydronium perchlorate according to (5.8). Disregarding the titrations of $Mg(OAc)_2$ and $Hg(OAc)_2$, all the enthalpy values lie between -7.2 kcal mol^{-1} for $Pb(OAc)_2$ and -5.7 kcal mol^{-1} for NaOAc when the reaction takes place in an anhydrous solvent. The order of magnitude of these heats of neutralization agrees with that expected by Kolthoff and Bruckenstein[32] from the temperature dependence of ion-pair formation. Titrations with hydronium perchlorate have enthalpy values that are on an average 2.7 kcal mol^{-1} higher than those with anhydrous perchloric acid.

The authors attribute this higher reaction enthalpy to the reaction

$$CH_3COOH_2^+ + H_2O \rightleftharpoons CH_3COOH + H_3O^+, \Delta H_R = -2.7 \text{ kcal mol}^{-1}$$

which is defined on the basis of the heat balance of (5.7) and (5.8). The differently shaped titration curves of the individual acetates indicate a complex system of reaction steps. There is not enough quantitative information available to develop a detailed theory, as has been done for aqueous solutions. The accuracy of the end-point indication of acetate titrations is very good, with a relative standard deviation of 0.1 to 0.5%.

A second group of experiments by the same authors deals with the thermometric titration of organic bases with perchloric acid under the same experimental conditions. The standard deviations of the end-point indication are likewise good. Even very weak bases such as urea or acetamide, which cannot be titrated by the potentiometric method, show sharp end points in thermometric titration. A comparison of the data for the enthalpies again shows a difference of about 2.7 kcal mol^{-1} between titration with anhydrous perchloric acid and with hydronium perchlorate. There seems to be no correlation of the heat of reaction and the strength of the base of the titrand in acetic acid.[37] Phosphates and nitrates and also salts of weak organic acids (sodium benzoate, sodium citrate, and potassium biphthalate) can be titrated with anhydrous perchloric acid in acetic acid.[36]

For a number of reasons, acetonitrile is a good solvent for titration reactions. It is a very weak base with practically no acidic properties and can, therefore, be used over a wide range of acidity. Its disadvantage is that hydroxyl ions catalyze its polymerization.

Forman and Hume[38] examined the usefulness of acetonitrile as a solvent in thermometric acid-base titrations. Nonaromatic, aromatic, and heterocyclic amines can be titrated accurately with hydrogen bromide solution; p-bromoaniline and m-chloroaniline represent the limit beyond which thermometric titration is not feasible. The pK_B value in aqueous solution can be used as a measure of the strengths of the bases. The weakest bases that can be titrated have a pK_B value of 10 to 10.5. Mixtures, such as n-butylamine and pyridine, yield two sharp end points. The heats of neutralization of the amines, which can be determined from the initial slope of the thermograms, provide a measure of the strength of aromatic, aliphatic primary, and the straight-chain secondary amines in acetonitrile.[39]

Acids can be titrated in acetonitrile with 1,3-diphenylguanidine.[38] The weakest acid that can still be titrated is benzoic acid (pK_a in water 4.20). Acetic acid and p-toluic acid show no hint of an end point. The heats of reaction of acid titration with 1,3-diphenylguanidine fall between -12.4 kcal mol^{-1} (for benzoic acid) and -25.3 kcal mol^{-1} (for hydrogen chloride).

Belisle[40] has shown that thallium(I)ethoxide is a suitable titrant for acids (e.g., m-methoxybenzoic and benzoic acids, p-nitrophenol, phenol, p-methoxyphenol, and trifluoroethanol) in aprotic solvents (benzene, carbon tetrachloride, chlorobenzene, and methylene chloride). Also, TlOC$_2$H$_5$ is soluble in the solvents mentioned. The thermograms of weak acids show sharp end points; only in the case of the very weak acid trifluoroethanol ($pK_a = 12.4$) is it necessary to obtain the end point by extrapolation.

Other investigations in this field deserving mention are the titration of bromo derivatives of phenols and cresols in alcoholic solutions with alcoholic KOH solution,[41] and that of phenols in pyridine.[42] To determine the acidity of petroleum products, Quilty[43] dissolved them in isopropanol and titrates with isopropanolic KOH.

The titrations discussed in the following section belong at least in part to the group of titrations of Brönsted acids and bases. They are discussed in a section of their own because of the special technique that is used for the end-point indication.

5.5 SPECIAL END-POINT INDICATION TECHNIQUES

5.5.1 CATALYTIC THERMOMETRIC TITRATION

In the search for a suitable nonaqueous solvent for the titration of acids with alkali, Vaughan and Swithenbank found acetone.[44]

A marked temperature rise occurs after the equivalence point with the first excess of a nonaqueous alcoholic alkali solution in the titration of an acid in anhydrous acetone. Even traces of water will drastically reduce this rise; it is hardly discernible if the water content is 2% or more. The addition of up to 25% (volume) of anhydrous solvents, such as pyridine, benzene, or nitrobenzene, reduces the slope of the curve after the equivalence point, but titration is still possible. If the alkali solution is added to pure acetone, the temperature rise occurs at the first addition. The reason for this phenomenon is a base-catalyzed reaction of the ketones forming β-ketols and higher products of condensation. The reaction has a heat effect of more than 800 kcal mol^{-1} KOH.

Vaughan and Swithenbank call acetone an "enthalpimetric indicator." Vajgand and Gaál[45] coined the term *catalymetric thermometric titration* for this type of titration. Catalytic thermometric titration reactions may be represented as[46]

$$A + B \rightarrow AB \quad \text{during the reaction period}$$
$$B$$
$$C + D \rightarrow P \quad \text{during the dilution period}$$

The titrant B is the reactant of the titration reaction. Under normal conditions in the range $t > t_\alpha$ it would produce the heat of dilution H_v if the titrant were added in excess. In a catalytic thermometric titration, on the other hand, it catalyzes the indicator reaction. The titration reaction must be fast and complete so that at no time during the reaction period does a high-enough concentration of B exist in the solution to start the indicator reaction.

Mono- and divalent carboxylic acids, hydroxy acids, mono- and polyhydric phenols, phenolic acids, keto-enols, and imides have been successfully titrated in acetone in this way.[44] As there is strong ionization in acetone,[47] titration of very weak acids, such as 2,6-disubstituted phenols, acetylacetone, and succinimide, becomes feasible.[44]

The catalyzed hydrolysis of acetic anhydride, which is the basis of the method of Greathouse et al.[8] is the starting point for another catalytic thermometric titration:[48] The base to be titrated is dissolved in glacial acetic acid, to which a small amount of water is added (2% of water together with 8% of acetic anhydride); a solution of perchloric acid in glacial acetic acid is used for titration. As long as the perchloric acid is consumed in the titration reaction, the hydrolysis reaction of the acetic anhydride can be neglected. After the equivalence point has been reached, the catalytic reaction begins and produces a sharp break and a steep rise of the titration curve. The method has been successfully used for the titration of tertiary amines and organic salts (Table 5.1). In a further paper,[49] nitromethane (with an addition of 2% water and 8% acetic anhydride) and acetic anhydride (2% water) have been reported as solvents for the catalytic thermometric titration. Weak bases, such as antipyrine (K_B in water $4 \cdot 10^{-13}$), which cannot be titrated by catalytic thermometric titration in acetic acid, can be titrated in the solvents mentioned above. The same paper reports that the discontinuous addition can be replaced by continuous addition in conjunction with automatic recording of the temperature. This suggests that continuous addition of the titrant may be replaceable by coulometric generation of hydrogen ions.[49,50,51]

TABLE 5.1. Results Obtained by Catalytic Thermometric Titration in Glacial Acetic Acid[48]

Substance	Theoretical g	Observed Experimental g	Number of Titration Replicates	Standard Deviation %
Aminopyrine	0.1500	0.1495	6	0.29
Antipyrine	0.2000	0.1942	8	0.11
Cinchonine	0.1000	0.0970	9	0.21
β-Hydroxyethyl-2-methyl-5-nitroimidazole	0.2000	0.1985	5	0.42
1,3-Triethyl ammonium-2-propanol iodide	0.2000	0.1999	6	0.06
p-Benzoquinone-amidinohydrazone-thiosemicarbazone	0.2000	0.1858	10	0.08
Potassium biphthalate	0.2000	0.2016	3	0.21
Sodium acetate	0.1500	0.1501	6	0.23
Sodium benzoate	0.1500	0.1496	6	0.03
Sodium formate	0.1000	0.0990	6	0.17
Sodium salicylate	0.2000	0.2007	6	0.11
Triethylamine	0.1500	0.1503	6	0.07
5-Hydroxy-6-methyl-3,4-pyridine dimethanol -HCl	0.2000	0.2016	9	0.21

Greenhow[52,53] and Greenhow and Spencer[54-56] have used catalytic thermometric procedures in which highly exothermic, ionic polymerization processes indicate the end point in titrations of organic acids and bases in nonaqueous solution. Acrylonitrile, used as the solvent in the titration of organic acids (benzoic acid and its derivatives; phenols; imides; and polyfunctional acids, such as succinic acid, pyrogallol, hydroxybenzoic acids, and cyanuric acids) with a strongly basic titrant (potassium hydroxide in isopropanol or tetra-n-butylammonium hydroxide in toluene–methanol), is polymerized by the catalytic action of the first excess of alkaline titrant after the acid has been neutralized.[54]

Similarly, α-methylstyrene can be used as the monomer component of the solvent in titrations with perchloric acid of basic compounds, such as aliphatic amines, pyridine and aniline derivatives, difunctional aromatic amines, heterocyclic nitrogen compounds, amides, and sulfur and phosphorus derivatives. The same titrations have been investigated in the monomer solvent isobutyl vinyl ether with boron trifluoride as the titrant.[55] The applicability of the method at the 10-μg level offers the possibility of pharmaceutical and forensic analysis. Small quantities of alkaloids, purine bases, and alkaloid salts have been successfully titrated with perchloric acid using α-methyl styrene as the indicator compound;[56] 10 μg of alkaloid can be determined with 0.001 M titrant at a precision of about 1%.

Catalytic thermometric methods have been published recently in which the function of indicator is assumed by a reaction prepared in the sample instead of by the solvent.[57] In the chelating reaction of EDTA, DCTA, and NTA (see Section 8.3.3) with Mn(II) ions, the indicator reaction is provided by H_2O_2 decomposition or the reaction of H_2O_2 with resorcinol, both catalyzed by manganese.[58] The end-point indication is strongly accentuated in DMSO and allows titrations with 10^{-5} M Mn(II) solution, which cannot be conducted at this concentration in water; the titration of NTA is possible in DMSO but not in water. By back titration of excess EDTA or DCTA solution with a standard Mn(II) nitrate solution, various ions can be determined in DMSO; it should be mentioned that a number of these determinations, which are possible in DMSO, cannot be performed in aqueous solution.

Catalytic thermometric titration is the most recent development of the thermometric method. In the hands of the analytical chemist it is an efficient tool that has already provided an answer to many hitherto unsolved titration problems.

5.5.2 END-POINT INDICATION WITH THE HELP OF A LARGE HEAT OF DILUTION

In this context another method that makes use of the properties of the system excess titrant–solvent can be mentioned. Vaughan and Swithenbank[59] observed a large endothermic heat of dilution when a strong hydrogen chloride solution in isopropyl alcohol is added to a wide range of organic solvents, excluding alcohols. In an acid-base titration a titrant of 5 N HCl in isopropyl alcohol first gives the expected temperature rise as a result of the neutralization of the base, followed by a sharp temperature drop that marks the end point. Aliphatic and aromatic amines, and the amides urea and thiourea were titratable in suitable solvents using this method. Other amides, indole, carbazole, anilides, and o-nitraniline could not be titrated. In some solvents, excepting acetic acid, mixtures of aliphatic and aromatic bases can be titrated by 5 N HCl in isopropyl alcohol with individual end points; the aliphatic bases are titrated first.

5.6 FURTHER TITRATIONS IN NONAQUEOUS SOLVENTS

A number of thermometric titrations cannot be classified in one of the preceding sections of this chapter and are discussed in the following paragraphs.

5.6.1 UNSATURATED COMPOUNDS

The thermometric titration method has been combined with the preferred method of olefin determination, that of hydrogenation.[12,60] The hydrogenation reaction is conducted as a DIE procedure in hexane solvent and with palladium catalyst on charcoal. At constant hydrogen pressure in the reaction chamber, between 1 and 20 μl of the olefin solution (20% olefin in hexane) is injected; the hydrogenation reaction is quantitative and rapid for most simple olefins. Within a working range of about 2 to 20 μmol, the authors obtained results with a relative standard deviation of 1 to 3%.

The determination of benzene in cyclohexane and of toluene in methylcyclohexane was achieved with nitrating acids,[61] and the most favorable conditions for this titration were studied.[62]

Jordan et al.[63] (see also Ref. 64) have determined olefinic unsaturation with iodine monochloride (see Section 7.5); the oldest work in this field is that of Somiya[65] on the determination of the iodine value of unsaturated fats based on the usual halogen reaction.

5.6.2 ACETYLATION

The thermometric determination of acetic acid in acetylating mixtures utilises the acid-catalyzed reaction of acetic acid with water (see Section 5.1) or is based on the strongly exothermic reaction between acetic acid and aniline.[66-70] In acetylating mixtures containing sulfuric acid, the latter can be titrated with barium acetate or lead acetate dissolved in glacial acetic acid.[68,71]

Snelson, Ellis, and Vilkauls[72] determine the hydroxyl value of alkylphenols with a routine method by direct injection enthalpimetry, using an acetylation mixture (ethyl acetate, acetic anhydride, and perchloric acid) into which the alkylphenol is injected.

Further application of thermometric titration was successful in acetylation of cellulose[70,73,74] and in the estimation of the acetyl value of oils and fats.[65,75]

5.7 THERMOMETRIC TITRATION IN MOLTEN SALTS

The investigation by Jordan et al.[76–79] of titration in molten salts furnishes a further example of the wide field of application of thermometric titration. The temperature of titration is in the range of 150 to 200°C; more or less standard equipment is used with automatic addition of titrant and a thermistor as temperature sensor. Four adiabatic measuring cells, 250-ml Dewar flasks, are fixed on a revolving disk in the thermostated stove and can thus be used in succession. The titrant also is thermostated. A thermostating system with random temperature fluctuations of no more than $\pm 0.0005°C$ is essential because the titration produces only a very small temperature rise. The content of KCl (samples $8 \cdot 10^{-4}$ to $2 \cdot 10^{-2}$ m prepared by weight) was determined in a eutectic melt $LiNO_3$–KNO_3 (158°C) using $AgNO_3$.[76] The precipitation titration $Ag^+ + Cl^- \rightarrow AgCl(s)$ produces the thermogram of an incomplete titration reaction. Silver chloride is about 20% soluble under the given experimental conditions. The chloride content of the titrand can be determined with an accuracy of 1% (standard deviation 2%).

An evaluation of the thermogram for thermodynamic data yields the solubility product of silver chloride: at 158°C: $K_{sp}^{158°} = m_{Ag^+} \cdot m_{Cl^-} = 3 \cdot 10^{-8}$ (mol per 1000 g solvent)^2 and its heat of precipitation $\Delta H_{AgCl, 158°} = -18.9 \pm 0.3$ kcal mol^{-1}. The evaluation procedures correspond to those discussed for aqueous solutions. End-point indication in the titration of KBr, KI, and K_2CrO_4 is accurate to 1% for halides and 4% for chromates.[77] Table 5.2 contains a synopsis of thermodynamic information obtained from the thermograms.

Metzger, Brenner, and Salmon[80] use the example of a KCl–CdCl$_2$ melt at 600 and 780°C to describe a method for the determination of the percentage composition of a complex and its stability constant from the partial molal heat. A small sample of each salt is added to a series of mixtures (0 to 100%) of the reactants. The evaluation shows a 1:1 compound with a stability constant of 3.3.

TABLE 5.2. Synopsis of Thermodynamic Information Obtained from $AgNO_3$ Titrations[77] at 431°K

Reaction	ΔG^0 kcal mol^{-1}	ΔH^0 kcal mol^{-1}	ΔS^0 cal mol^{-1} deg^{-1}
$Ag^+ + Cl^- \rightleftharpoons AgCl(s)$	−18.9	−14.9	−9.3
$Ag^+ + Br^- \rightleftharpoons AgBr(s)$	−26.1	−17.7	−19.5
$Ag^+ + I^- \rightleftharpoons AgI(s)$	−32.1	−24.2	−18.3
$2Ag^+ + CrO_4^{2-}(s) \rightleftharpoons Ag_2CrO_4(s)$	−16.7	−15.7	−2

REFERENCES

1. T. Higuchi and C. R. Rehm, *Anal. Chem.*, **27**, 409 (1955).
2. R. J. N. Harries, *Talanta*, **15**, 1345 (1968).
3. J. J. Lagowski, *Anal. Chem.*, **42**, 305R (1970).
4. K. Fischer, *Angew. Chem.*, **48**, 394 (1935).
5. J. C. Wasilewski and C. D. Miller, *Anal. Chem.*, **38**, 1750 (1966).
6. L. Erdey and J. Maric, *Magy. Kémikusok Lapja*, **25**, 584 (1970).
7. M. Y. Spink and C. H. Spink, *Anal. Chem.*, **40**, 617 (1968).
8. L. H. Greathouse, H. J. Janssen, and C. H. Haydel, *Anal. Chem.*, **28**, 357 (1956).
9. R. Szabó, *Z. Phys. Chem.*, **122**, 405 (1926).
10. L. S. Bark and P. Bate, *Analyst*, **97**, 783 (1972).
11. J. H. McClure, T. M. Roder, and R. H. Kinsey, *Anal. Chem.*, **27**, 1599 (1955).
12. D. W. Rogers, *Anal. Chem.*, **43**, 1468 (1971).
13. T. R. Crompton and B. Cope, *Anal. Chem.*, **40**, 274 (1968).
14. J. Barthel and K. Wachter-Lenz, unpublished.
15. Y. Trambouze, *C. R. Acad. Sci. (Paris)*, **233**, 648 (1951).
16. Y. Trambouze, L. de Mourgues, and M. Perrin, *C. R. Acad. Sci. (Paris)*, **234**, 1770 (1952).
17. I. P. Romm, E. N. Gur'yanova, I. P. Gol'dshtein, and K. A. Kocheshkov, *Dokl. Akad. Nauk SSSR*, **172**, 618 (1967).
18. S. T. Zenchelsky, J. Periale, and J. C. Cobb, *Anal. Chem.*, **28**, 67 (1956).
19. S. T. Zenchelsky and P. Segatto, Abstr. of Papers, 132nd Meeting Am. Chem. Soc., New York, 1957, p. 12B.
20. S. T. Zenchelsky and P. R. Segatto, *J. Am. Chem. Soc.*, **80**, 4796 (1958).
21. F. J. Cioffi and S. T. Zenchelsky, *J. Phys. Chem.*, **67**, 357 (1963).
22. I. P. Gol'dshtein, E. N. Gur'yanova, and I. R. Karpovich, *Russ. J. Phys. Chem.*, **39**, 491 (1965).
23. I. P. Gol'dshtein, E. N. Gur'yanova, and K. A. Kocheshkov, *Dokl. Akad. Nauk SSSR*, **161**, 111 (1965).
24. I. P. Gol'dshtein, E. N. Gur'yanova, N. N. Zemlyanskii, O. P. Syutkina, E. M.

Panov, and K. A. Kocheshkov, *Dokl. Akad. Nauk SSSR*, **175**, 836 (1967).
25. B. Nelander, *Acta Chem. Scand.*, **20**, 2289 (1966).
26. R. S. Drago, T. F. Bolles, and R. J. Niedzielski, *J. Am. Chem. Soc.*, **88**, 2717 (1966).
27. E. G. Hoffmann and W. Tornau, *Z. Anal. Chem.*, **188**, 321 (1962).
28. W. L. Everson and E. M. Ramirez, *Anal. Chem.*, **37**, 806 (1965).
29. W. L. Everson and E. M. Ramirez, *Anal. Chem.*, **37**, 812 (1965).
30. W. L. Everson, *Anal. Chem.*, **36**, 854 (1964).
31. R. D. Parker and T. Vlismas, *Analyst*, **93**, 330 (1968).
32. I. M. Kolthoff and S. Bruckenstein, *J. Am. Chem. Soc.*, **78**, 1 (1956).
33. S. Bruckenstein and I. M. Kolthoff, *J. Am. Chem. Soc.*, **78**, 10 (1956).
34. S. Bruckenstein and I. M. Kolthoff, *J. Am. Chem. Soc.*, **78**, 2974 (1956).
35. W. Huber, *Titrationen in nichtwässrigen Lösungsmitteln*, Akademische Verlagsgesellschaft, Frankfurt am Main, 1964.
36. H. J. Keily and D. N. Hume, *Anal. Chem.*, **36**, 543 (1964).
37. N. F. Hall, *J. Am. Chem. Soc.*, **52**, 5115 (1930).
38. E. J. Forman and D. N. Hume, *Talanta*, **11**, 129 (1964).
39. E. J. Forman and D. N. Hume, *J. Phys. Chem.*, **63**, 1949 (1959).
40. J. Belisle, *Anal. Chim. Acta*, **54**, 156 (1971).
41. R. Pâris and J. Vial, *Chim. Anal.*, **34**, 3 (1952).
42. J. S. Parsons, Abstr. of Papers, 132nd Meeting, Am. Chem. Soc., New York, 1957, p. 12B.
43. C. J. Quilty, *Anal. Chem.*, **39**, 666 (1967).
44. G. A. Vaughan and J. J. Swithenbank, *Analyst*, **90**, 594 (1965).
45. V. J. Vajgand and F. F. Gaál, *Bull. Soc. Chim. Beogr.*, **31**, 103 (1966).
46. V. Vajgand, F. Gaál, Lj. Zrnić, S. Brusin, and D. Velimirović, *Proceedings of the 3rd Analytical Chemistry Conference*, 1970, 2, 443.
47. H. B. van der Heijde and E. A. M. F. Dahmen, *Anal. Chim. Acta*, **16**, 378 (1957).
48. V. J. Vajgand and F. F. Gaál, *Talanta*, **14**, 345 (1967).
49. V. J. Vajgand, T. A. Kiss, F. F. Gaál, and I. J. Zsigrai, *Talanta*, **15**, 699 (1968).
50. V. J. Vajgand and R. Mihajlović, *Talanta*, **16**, 1311 (1969).
51. V. J. Vajgand, F. F. Gaál, and S. S. Brusin, *Talanta*, **17**, 415 (1970).
52. E. J. Greenhow, *Chem. Ind.*, **1972**, 466.
53. E. J. Greenhow, *Chem. Ind.*, **1972**, 422.
54. E. J. Greenhow and L. E. Spencer, *Analyst*, **98**, 81 (1973).
55. E. J. Greenhow and L. E. Spencer, *Analyst*, **98**, 90 (1973).
56. E. J. Greenhow and L. E. Spencer, *Analyst*, **98**, 98 (1973).
57. T. Kiss, *Z. Anal. Chem.*, **252**, 12 (1970).
58. H. Weisz and T. Kiss, *Z. Anal. Chem.*, **249**, 302 (1970).
59. G. A. Vaughan and J. J. Swithenbank, *Analyst*, **92**, 364 (1967).
60. D. W. Rogers and R. J. Sasiela, *Talanta*, **20**, 232 (1973).
61. B. B. Corson and L. J. Brady, *Ind. Eng. Chem.*, Anal. Ed., **14**, 531 (1942).
62. R. L. Bishop and E. L. Wallace, *Ind. Eng. Chem.*, Anal. Ed., **15**, 563 (1943).
63. J. Jordan, P. T. -S. Pei, and R. C. Buchta, Jr., *Abstr. 149th Nat. Meeting, Am. Chem. Soc.*, Detroit, 1965, p. 30B.

64. J. Jordan, R. A. Henry, and J. C. Wasilewski, *Microchem. J.*, **10**, 260 (1966).
65. T. Somiya, *J. Soc. Chem. Ind. (Jap.)*, **33**, 174 (1930) (Suppl.).
66. H. D. Richmond and J. A. Eggleston, *Analyst*, **51**, 281 (1926).
67. T. Somiya, *Proc. Imp. Acad. (Jap.)*, **3**, 76 (1927).
68. T. Somiya, *Proc. Imp. Acad. (Jap.)*, **3**, 79 (1927).
69. T. Somiya, *Proc. Imp. Acad. (Jap.)*, **5**, 34 (1929).
70. T. Somiya, *J. Soc. Chem. Ind. (Jap.)*, **32**, 493 (1929).
71. T. Somiya, *J. Soc. Chem. Ind. (Jap.)*, **31**, 306 (1928).
72. F. L. Snelson, W. R. Ellis, and J. Vilkauls, *Analyst*, **92**, 264 (1967).
73. L. H. Greathouse, *Abstr. of Papers*, Sept. Meeting Am. Chem. Soc., 1957, p. 7B.
74. T. Somiya, *J. Soc. Chem. Ind. (Jap.)*, **32**, 153 (1929) (Suppl.).
75. T. Somiya, *J. Soc. Chem. Ind. (Jap.)*, **33**, 140 (1930) (Suppl.).
76. J. Jordan, J. Meier, E. J. Billingham, Jr., and J. Pendergrast, *Anal. Chem.*, **32**, 651 (1960).
77. J. Jordan, J. Meier, E. J. Billingham, Jr., and J. Pendergrast, *Nature*, **187**, 318 (1960).
78. J. Jordan, *Chimia*, **17**, 101 (1963).
79. J. Jordan, J. Meier, E. J. Billingham, Jr., and J. Pendergrast, *Anal. Chem.*, **31**, 1439 (1959).
80. W. H. Metzger, Jr., A. Brenner, and H. J. Salmon, *J. Electrochem. Soc.*, **114**, 131 (1967).

CHAPTER

6

THERMOMETRIC PRECIPITATION TITRATION

6.1 THEORETICAL BASIS OF PRECIPITATION TITRATION

If the heats of precipitation are sufficiently large, the thermometric method can be used for precipitation titrations with the same direct simplicity that we have seen in the preceding discussion of acid-base titrations. The thermometric method has already been suggested as a useful tool for end-point indication in precipitation titrations by Dutoit and Grobet,[1] Dean et al.,[2,3] and Mayr and Fisch.[4] End-point indication using it in complete titration reactions is as accurate as that in acid-base titrations, namely 0.2%. Tyson, McCurdy, and Bricker[5] quote an accuracy of 0.1% for the titration of $AgNO_3$ with HCl in a method of differential thermometric analysis.

The thermogram of a pure precipitation titration with complete titration reaction (Fig. 6.1) contains the heat of precipitation as the only contribution to the integral heat effect.

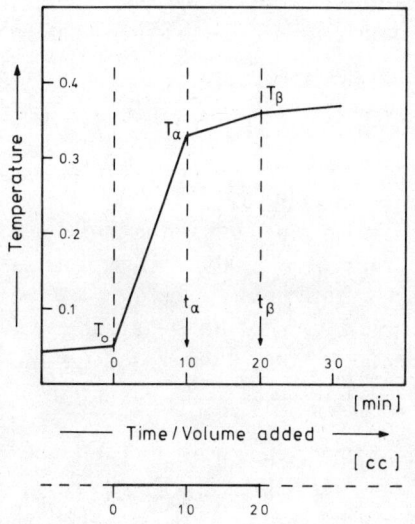

Fig. 6.1. Thermogram of the titration of $AgNO_3$ with KI ($V_0 = 200$ ml, $a = 0.00625$ M, $b = 0.125$ M) in methanol as solvent.[6]

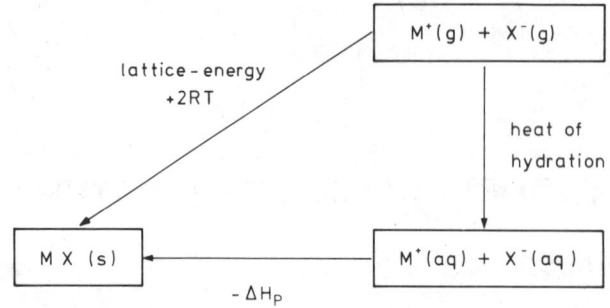

Fig. 6.2. Diagram illustrating the signification of ΔH_P in aqueous solutions.[6]

$$Q_R(t) = \int_0^t w_p(t)\,dt = -\Delta H_P b v_0' t \tag{6.1}$$

As $w_p(t)$ vanishes with the addition of an equivalent volume of $v_\alpha = v_0' t_\alpha$, a break occurs in the curve at the point t_α, and this is used for indicating the end point.

The significance of the enthalpy of precipitation for an example such as $M^+(aq) + X^-(aq) = MX(s)$ in aqueous solution is illustrated clearly in Fig. 6.2.[6]

$$\Delta H_p = \Delta H_{\text{hydration}} - \Delta H_{\text{lattice}} \tag{6.2}$$

Lattice energies[7,8] and heats of hydration[8] are known for silver halides so that (6.2) can be verified with the data obtained in the precipitation titration:[6]

$$\text{AgNO}_3(\text{solv}) + \frac{1}{n}\text{MX}_n(\text{solv}) \longrightarrow \text{AgX}(s) + \frac{1}{n}\text{M(NO}_3)_n(\text{solv})$$

The results of this investigation are given in Table 6.1.

The titrations of Table 6.1 have been so arranged that the end point could be expected after adding 10 ml titrant to 200 ml silver nitrate solution as titrand. The accuracy of the end point is defined as the deviation from this theoretical value. The titration results, the accuracy of the end point, and ΔH_P of Table 6.1 are mean values based on the evaluation of several thermograms. The deviations for ΔH_P are within a range of 0.2 kcal mol^{-1}.

The precipitation enthalpies depend on the concentration (Table 6.1). The dependence on ion concentration for silver halides $\Gamma = \Sigma c_i z_i^2$, according to the following equation,

$$\Delta H = \Delta H^0 - A\sqrt{\Gamma} \tag{6.3}$$

TABLE 6.1. Precipitation of Silver Halides[6]

Titrand 200 ml $AgNO_3$ mol liter^{-1}	Titrant	Concentration of Titrant mol liter^{-1}	Solvent	Accuracy of end point %	Results of Titration			
					ΔH_p kcal mol^{-1}	$\Delta H_{lattice}$ and Cited Reference kcal mol^{-1}	$\Delta H_{solvation}$ (6.2) kcal mol^{-1}	and Cited Reference kcal mol^{-1}
1. 0.1000	KBr	2.000	H_2O	0.2	−20.3	−214[8]	−194	−194[8]
0.0500		1.000			−20.2	−212[6]		
0.0250		0.500			−19.9			
0.0125		0.250			−19.9			
2. 0.0500	HCl	1.000	H_2O	0.2	−15.8	−216[8]	−200	−200[8]
	NaCl	1.000			−15.6	−215[6]		
	$BaCl_2$	0.500			−15.8			
	$LaCl_3$	0.333			−15.6			
3. 0.0500	KI	1.000	H_2O	0.2	−26.9	−211[8]	−184	−183[8]
						−211[6]		
4. 0.0125	KI	0.250	CH_3OH	0.2	−21.6	−211[8]	−189	Unknown
0.00625		0.125		0.3	−21.6	−211[6]		

was first investigated by Ewing and Mazac.[9] Table 6.1 (1.) gives a value of $A = 1.4$ kcal liter$^{1/2}$ mol$^{-3/2}$ against that of 8.5 kcal liter$^{1/2}$ mol$^{-3/2}$ given by Ewing. We must assume that there are irreproducible heat effects of the halide precipitation within this range that are not explained by the ion model from which (6.3) is derived. According to the Debye–Hückel theory (6.3), all halides should give the same value for A depending solely on the temperature of the solvent and its dielectric constant ($A \approx 0.4$ kcal liter$^{1/2}$ mol$^{-3/2}$ in water, $t = 25°C$). The heat of precipitation does not depend on the nature of the cation of the soluble halide [Table 6.1 (2.)].

If the lattice energy is known, the heat of precipitation derived from the thermogram and (6.2) gives an enthalpy of solvation that agrees with the enthalpy values[8] determined with other methods. Thus the thermometric method provides a simple way of determining enthalpies of solvation. This is of special interest for precipitation reactions in nonaqueous solvents [Table 6.1 (4.)]. According to Fig. 6.2 and (6.2), information can be gained corresponding to the process

$$M^+(aq) + X^-(aq) \rightarrow M^+(solv) + X^-(solv)$$

that is, the transfer of the ions from water into a different solvent. The value is +5.3 kcal mol^{-1} for the transfer of AgI from water into methanol. This energy cannot, however, be partitioned and attributed to single ions without studying the association of the salts in the various solutions.

The difficulty of obtaining precipitates of a defined structure is well known,[10] especially if the reaction is fast.[11] The adsorption of ions by the precipitation product presents an additional problem[11] (see Section 6.2.9). Effects of this kind result in a shift of the end point; their own thermal power w_i contributes to the integral heat effect observed, which can then no longer be described by (6.1).

$$Q_R(t) = \int_0^t w_P(t)\,dt + \sum_i \int_0^t w_i(t)\,dt \qquad (6.4)$$

Other possible sources, apart from adsorption, of supplementary reaction powers $w_i(t)$ are the processes of re-solution of the precipitation, complexation of ions, dissociation, and so forth.

Equation (6.4) has been studied experimentally in the precipitation of barium sulfate.[6] The results are presented in Table 6.2. Several cases can be shown in which this reaction produces w_i-type energies. The reaction Na$_2$SO$_4$ with Ba(NO$_3$)$_2$ and its reversal [Table 6.2 (1.)] produces the true heat of precipitation $\Delta H_P = -4.5$ kcal mol^{-1} for the precipitation of

TABLE 6.2. Precipitation of Barium Sulfate[6]

Titrand 200 ml	Concentration of titrand mol liter^{-1}	Titrant	Solvent	Accuracy of end point %	ΔH_R kcal mol^{-1}	ΔH_P kcal mol^{-1}	ΔH_D kcal mol^{-1}	ΔH_N kcal mol^{-1}
1. Na$_2$SO$_4$	0.050	Ba(NO$_3$)$_2$	H$_2$O	0.2	−4.55	−4.5		
Ba(NO$_3$)$_2$	0.050	Na$_2$SO$_4$		0.2	−4.49	−4.5		
	0.025	Na$_2$SO$_4$		0.2				
2. Ba(NO$_3$)$_2$	0.050	Na$_2$SO$_4$	3.0 M NaCl 0.5 M HNO$_3$	1.0 —				
3. Ba(NO$_3$)$_2$	0.050	H$_2$SO$_4$	H$_2$O	0.2	−9.42	(−4.5)	−5.3	
BaCl$_2$	0.050	H$_2$SO$_4$		0.5	−9.28	(−4.5)	−5.4	
H$_2$SO$_4$	0.015	Ba(NO$_3$)$_2$		0.3	−7.78	(−4.5)	−5.3	
4. Ba(OH)$_2$	0.025	H$_2$SO$_4$	H$_2$O	0.2	−34.58	(−4.5)	−5.4	(−13.3)

barium sulfate. According to (6.4), $\Sigma_i \int w_i(t) = 0$. High concentrations of foreign salts do not interfere appreciably in the reaction. The precipitation yields a diagram that can no longer be evaluated, however, if HNO_3 is present [Table 6.2 (2.)]. The titration of H_2SO_4 with $Ba(NO_3)_2$ produces the heat of dissociation of sulfuric acid $\Delta H_D^{(2)}$, together with the heats of precipitation and dilution, and gives

$$w_i(t) = -\Delta H_D^{(2)} a V_0 \frac{d\alpha^{(2)}}{dt}$$

where a is the concentration of the titrand, V_0 the volume of the sample, $\alpha^{(2)}$ is the degree of dissociation of the second proton of the sulfuric acid, which changes during the titration from $\alpha_0^{(2)}$ (beginning of titration) to $\alpha^{(2)} = 1$ (end of titration). The term $\alpha^{(2)}$ is calculated according to Section 1.3.3. Table 6.2 contains in brackets the known values introduced for the evaluation of the others. An evaluation of the heat of reaction gives for a value of $\Delta H_P = -4.5$ kcal mol^{-1} a heat of dissociation for the second proton of $\Delta H_D^{(2)} = -5.3 \pm 0.2$ kcal mol^{-1}. Christensen et al.[12] found a value of -5.4 kcal mol^{-1} for the dissociation of sulfuric acid.

When the barium salt is used as the titrand and sulfuric acid as the titrant, the reaction period and the period of dilution take a share of the heat of dissociation, and this must be taken into account in the equation used for evaluation. Identical $\Delta H_D^{(2)}$ values are then yielded, in spite of the considerable differences in the ΔH_R values [Table 6.2 (3.)]. When $Ba(OH)_2$ is the titrand, twice the heat of neutralization must be added for the reaction period. The titrations mentioned so far all belong to a group characterized by a sharp break in the thermogram. Hence at least one of the integral functions in (6.4) disappears at the point $t = t_\alpha$. In halide precipitation or precipitation of sulfate, $Na_2SO_4 + Ba(NO_3)_2$, ΔH_P disappears and with it the entire heat of reaction; in the precipitation of sulfate, $Ba(OH)_2 + H_2SO_4$, the heats of precipitation and neutralization disappear, whereas the heat of dissociation of sulfuric acid furnishes a considerable share of the heat of reaction beyond $t = t_\alpha$.

A thermogram of the most general form has no sharp breaks. Every heat-producing process contributes its share beyond the stoichiometric equivalence point, depending on the equilibrium of precipitation. Thermograms of this general type can be evaluated to yield equilibrium constants, heats of formation, and so forth (see Chapter 8). As a rule they are not very useful for determining the titer of the titrand because this requires too complicated and time-consuming calculations.

Depending on the position of the equilibrium of precipitation, many transitional types of diagrams are found, from the most general form to that with a sharp break at t_α. If the equilibrium lies toward the reaction products, construction of the end point is possible on the basis of adjoining sections of the curve. An example is the titration[6] of $Pb(NO_3)_2$ with H_2SO_4. This is an incomplete titration reaction, but the end point can be constructed from the adjoining sections of the curve with an accuracy of about 5%. Then ΔH_P is calculated from the heat of reaction as 1.3 ± 0.3 kcal mol^{-1}.

In precipitation reactions of this kind, the DIE method[13] is often the most suitable, since there is an excess of titrant that leads to a practically complete reaction.

6.2 SPECIAL TYPES OF PRECIPITATION REACTION

No borderlines can be drawn between the classes precipitation titration, complexometric titration, and titration in nonaqueous solutions. In the course of a complex-forming reaction, precipitation can occur in one of the reaction steps or the precipitate can be dissolved in another. In Lewis acid titration, the precipitation of reaction products is frequently observed. Some aspects of precipitation titrations are therefore discussed in Chapters 5 and 8, and these chapters should also be consulted in connection with this subject.

6.2.1 HALIDES AND CYANIDES

The method for the determination of chloride and cyanide ions proposed by Dean and Newcomber[3] was questioned by Mayr and Fisch.[4] The possibility of determining halides and cyanides by thermometric indication in a precipitation reaction was demonstrated several times in later investigations. Apart from the papers mentioned above in connection with this problem,[5,6,9] reference must be made to Refs. 14 and 15. Tyrrell and Beezer[16] report that Harries investigated the precipitation of various halides from solutions in which they are present in ratios from 1:4 to 4:1. The solubility product L and the heat of precipitation ΔH_P (for AgI $L = 1.5 \cdot 10^{-16}$, $\Delta H_P = 26.8$ kcal mol^{-1}; for AgBr $L = 7.7 \cdot 10^{-13}$, $\Delta H_P = 20.4$ kcal mol^{-1}; and for AgCl $L = 1.56 \cdot 10^{-10}$ and $\Delta H_P = 15.8$ kcal mol^{-1}) indicate that precipitation should take place in the following sequence: AgI—AgBr—AgCl, and that the slopes of the titration curves in the different sections should differ considerably. The end points are accurate to 1 to 5%, depending on the ratio in the mixture. A mixture of iodide and thiocyanate can also be titrated with silver nitrate and gives two sharp end points.[17]

The precipitation of Hg^{2+} as HgI_2 ($\Delta H_P^0 = -25$ kcal mol^{-1}) yields a diagram similar to Fig. 6.3. The end point is accurate to 1%; the complexation to HgI_4^{2-}, which follows the precipitation period, cannot be used for analytical purposes.[6]

Fluoride can be titrated through the precipitation reaction[18] $Pb^{2+} + F^- + Cl^- \rightarrow PbFCl$ ($\Delta H_P^0 = -36.6$ kJ mol^{-1}); in analogous reactions lead can be determined as PbF_2 and $PbNO_3F$, tin as $SnFCl$;[19] further results of Deschamps et al.[19] and Everson and Ramirez[20] on the determination of fluoride are discussed in Section 8.3.1.

According to Harries (quoted in Ref. 16), the titration of $AgNO_3$ with KCN produces two end points, one for $Ag(CN)_2^-$ and one for AgCN. On the other hand, Rasmussen and Nielsen[21] found in the thermogram two sections of differing slope, but they could not localize an end point for AgCN. If NH_3 is added, the end point for $Ag(CN)_2^-$ is obtained with an accuracy of 0.3%. The titration of Ni gives similar results;[21] when NH_3 is added, the end point for $Ni(CN)_4^{2-}$ is obtained with excellent sharpness. Nickel cannot be titrated in the presence of Ag; only the end point of the entire reaction can be obtained. The titration of Hg with KCN is reported to be unsatisfactory by Rasmussen and Nielsen.[21] Mondain-Monval and Pâris,[22] however, state that the homogeneous reaction of $HgCl_2$ and KCN produces a thermogram in which the end points for $Hg(CN)_2$ and $K_2Hg(CN)_4$ can be clearly recognized (see Fig. 3 in Ref. 22). The same authors have also studied other precipitation titrations with KCN. The titration of Ni yielded recognizable end points for $Ni(CN)_2$ and $K_2Ni(CN)_4$; that of Zn, end points for $Zn(CN)_2$ and $K_2Zn(CN)_4$; and that of Co, the end points[23] $Co(CN)_2$ and $Co(CN)_2 + 3KCN$.

Pâris investigated thermometric titration reactions in which insoluble ferrocyanides $Fe(CN)_6^{4-}$ are formed.[24] The titrations of Pb^{2+}, Ag^+, Zn^{2+}, Fe^{3+}, Ni^{2+}, Cu^{2+}, Co^{2+}, and Cd^{2+} with $K_4Fe(CN)_6$ produced diagrams with an indication of more than one end point for precipitates that generally contained potassium. Thus the titration of $CuSO_4$ gives $Cu_3K_2[Fe(CN)_6]_2$; the reverse titration, $K_4Fe(CN)_6$ with $CuSO_4$, on the other hand, produces $Cu_4K_4[Fe(CN)_6]_3$.

6.2.2 FLUOROCOMPLEX IONS

Aluminum, after reaction to K_3AlF_6 with hydrofluoric acid and potassium chloride, can be precipitated as NaK_2AlF_6. This precipitation reaction serves both for the determination of aluminum[25-27] and sodium;[28] Ca precipitates[27] as $KCaAlF_6$. These aluminum determinations, including the determination as $H_3(AlF_6)$ (see Section 8.3.1), and the determination of silicon as $H_2(SiF_6)$ or $K_2(SiF_6)$ (see Section 8.3.1), are part of the important work of Sajó et al. in the industrial analysis of steel, slags, clays, and minerals by their direct-reading method (see Section 3.2.3).

Everson,[29] who determined aluminum in an analogous way, observed that Be, Zr and Th interfere in addition to Ca, the only common metal that interferes at low concentrations and that cannot be masked; about 25 other metals, including Ti, Fe, and the lanthanides, either do not interfere or can be masked.

The thermometric method based on the reaction $2K^+ + SiF_6^{2-} \rightarrow K_2SiF_6$ is a good substitute for the gravimetric method of determining potassium;[30] Ba^{2+}, Sr^{2+}, Ca^{2+}, and Al^{3+} should not be present during this titration, whereas Fe^{3+}, Pb^{2+}, Li^+, Mg^{2+}, Mn^{2+}, Ni^{2+}, Zn^{2+}, and Cu^{2+} do not interfere in the reaction.

Further information about fluorocomplex titrations, both by precipitation and complexation reactions, are found in Refs. 19 and 20 (see Section 8.3.1).

6.2.3 HYDROXIDES AND BASIC SALTS

Dutoit et Grobet[1] studied the formation of basic salts of various metals by thermometric titration and found inflections of the titration curves that are utilizable for quantitative determinations, for example, $Pb(OH)_2$ (see Ref. 6).

The titration of Cd^{2+} and Zn^{2+} produces breaks at the points $1Cd^{2+} : 2OH^-$ and $1Zn^{2+} : 2OH^-$, respectively.[31] Tin as $SnCl_2$ (0.02 M) in HCl solution of known concentration can also be titrated in oxygen-free inert gas with NaOH, yielding $Sn(OH)_2$.[6]

Figure 6.3 is the diagram of a multistep reaction,[6] the titration of aluminum in a $KAl(SO_4)_2$ solution. The first reaction period, the period of precipitation $Al^{3+} + 3OH^- \rightarrow Al(OH)_3$, is followed after a sharp break at t_α by a second period in which the precipitate dissolves according to $Al(OH)_3 + OH^- \rightarrow [Al(OH)_4]^-$. This period continues without a break until the period of dilution. The point t_α can be used for determining the aluminum content of the sample. An end point $t_{\alpha 1}$ for the conversion into $[Al(OH)_4]^-$, constructed through extrapolation, is spurious because the corresponding reaction requires an excess of OH^-.

6.2.4 TETRAPHENYLBORATES

The search for reactants for K^+ led to successful titration with sodium tetraphenylborate, added in excess and subsequently back-titrated with standard KCl solution (K^+ in $N/100$ solution; accuracy 1%). Because they form insoluble tetraphenylborates,[30] NH_4^+, Ag^+, Tl^+, and Hg^{2+} interfere. A recent investigation[32] considers the conditions under which a direct thermometric titration of K^+, NH_4^+, and Tl^+ is possible. Errors of 1%

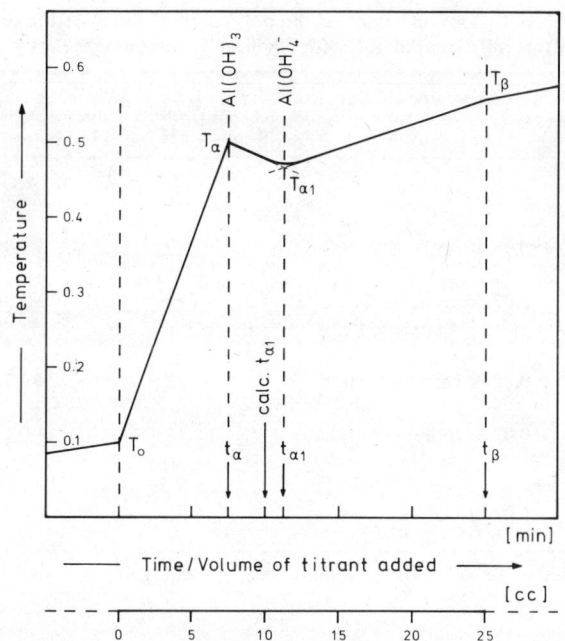

Fig. 6.3. Thermogram of the titration of $KAl(SO_4)_2 \cdot 12H_2O$ with NaOH ($V_0 = 200$ ml, $a = 0.025\ M$, $b = 2.0\ M$).[6]

arise for K^+ only at concentrations of less than 0.01 M as a result of its solubility product. The advantages of the thermometric method in comparison with possible conductivity or potentiometric titration are, of course, that it is independent of ionic strength and pH value of the solution (see Table 6.3). The enthalpy of this precipitation reaction is known from a further publication.[33]

Errors and uncertainty at pH = 1 result from the instability of the tetraphenylborate ion toward acids [$B(C_6H_5)_4^- + H^+ \rightarrow B(C_6H_5)_3 + C_6H_6$; $B(C_6H_5)_3 + 2H_2O \rightarrow (C_6H_5)B(OH)_2 + 2C_6H_6$]. Titrations in 1 N HCl produce deviations greater than 20%.

Bark and Grime[34] apply $B(C_6H_5)_4^-$ successfully in the titration of nitrogen-containing bases. The nature of a base $R_1R_2R_3R_4N^+$ ($R_i =$ hydrogen, alkyl, aryl; or the compound can be a heterocyclic base) predetermines the state of the precipitation reaction $R_1R_2R_3R_4N^+ + B(C_6H_5)_4^- \rightarrow R_1R_2R_3R_4NB(C_6H_5)_4$ after a given time; however, a suitable excess of titrant ensures that the reaction is completed in a standard time. The excess of tetraphenylborate is then determined by addition of a fixed volume of KCl solution in a DIE titration.

TABLE 6.3. The Influence of the pH Value of the Solution on the Determination of K^+ with Sodium Tetraphenylborate[32]

Sample	Solvent	Approximate pH	Error[a] %	σ %
KCl 0.029 M	0.1 M HCl	1	+2.1	2.4
	0.01 M HCl	2	+0.9	0.2
	0.07 M acetic acid–acetate buffer	5	+1.3	0.3
	0.07 M calcium acetate	8	+0.7	0.2
	0.07 M sodium phosphate buffer	12	+0.1	0.4
	0.07 M NaOH	13	+0.7	0.2

[a] The error refers to the deviation from the known amount of the sample.

6.2.5 PERCHLORATES

A DIE method for the titration of perchlorate ion has been worked out by Carr and Jordan.[35] It is based on the precipitation of ClO_4^- ions as tetraphenylarsonium perchlorate (solubility product $1.2 \cdot 10^{-8}$) in the direct reaction $ClO_4^- + As(C_6H_5)_4^+ \rightarrow As(C_6H_5)_4ClO_4$ ($\Delta H^0 = -10.9$ kcal mol^{-1}). A titration with $P(C_6H_5)_4^+$ ($\Delta H^0 = -10.5$ kcal mol^{-1}) produces equally good results; that with $Sb(C_6H_5)_4^+$ produces precipitation in a clearly time-dependent reaction (solubility product of the perchlorate $3.5 \cdot 10^{-8}$; $\Delta H^0 = -11.5$ kcal mol^{-1}). Permanganate, chlorate, and fluoroborate ions interfere.

6.2.6 OXALATES

Comprehensive material has been published on the titration of oxalates. Mayr and Fisch[4] give a synopsis of the results obtained by the classical gravimetric method and the thermometric method for the precipitation of Ca^{2+}, Sr^{2+}, Ba^{2+}, and Hg_2^{2+}, using ammonium oxalate as titrant. Mixtures of Hg_2^{2+} and Hg^{2+} yield two end points in the thermogram, both of which can be evaluated; the method is less suitable for Ba^{2+}.

The analysis of limestone (methods are discussed in Ref. 36), especially the simultaneous titration of Ca^{2+} and Mg^{2+}, is of technical interest. Chatterji[37] and Jordan and Billingham[38] report their precipitation as oxalates. Ammonium oxalate solution immediately precipitates Ca^{2+} in borate buffer solution (pH = 8) as $CaC_2O_4 \cdot H_2O$ ($\Delta H^0 = 6.1$ kcal mol^{-1}). This titration reaction produces a sharp end point. With Mg^{2+} as a sample under the same reaction conditions, neither precipitation nor rise of temperature could be observed even 30 min after addition of the ammonium oxalate solution. The precipitation of the magnesium oxalate is kinetically inhibited by a slow reaction during the stage of nuclei formation.[39] This kinetic effect can be used for the titration of Ca^{2+} in the presence of Mg^{2+} as long as the proportion of Ca^{2+} to Mg^{2+} in the sample does not exceed 1:2. In an analysis of dolomite or limestone the accompanying substances Al_2O_3 and Fe_2O_3 do not interfere. Aluminum and iron are precipitated as hydroxides during a fusion of the sample at a pH < 8. The critical proportion of 1 Ca^{2+} to 2 Mg^{2+} is not reached, even in the dolomite with its larger magnesium content.[38]

The precipitation of calcium oxalate for determining the calcium content of slags, cements, industrial silicates, and minerals is included in the thermometric analysis methods used by Sajó et al.[26,27,40–42]

6.2.7 PHOSPHATES

This precipitation reaction for magnesium is also a part of the direct-injection method of Sajó et al.[25–27,40,41]; diammonium hydrogen phosphate is used as a titrant. Titration with $NaNH_4HPO_4$ was used by Chatterji to determine magnesium in dolomite.[37]

The determination of phosphorus is conducted by the precipitation of phosphate with ammonium molybdate as phosphomolybdate [$(NH_4)_3PO_4 \cdot 12MoO_3$], which is determined by indirect titration (see Section 3.4.2, and Refs. 40 and 43).

6.2.8 CARBONATES

Pb^{2+}, Ag^+, Zn^{2+}, Cu^{2+}, and Al^{3+} can be precipitated with 1 M Na_2CO_3. Analyses have been made of samples in the range $M/30$ to $M/90$, with a standard deviation of about 1%.[15]

6.2.9 SULFATES

Sulfate precipitation served in Section 6.1 as a model to study thermometric precipitation titrations. It is used in analytical practice for both barium and sulfate determination.[2,4,6,11,40–42,44,45] The precipitation of lead as $PbSO_4$ is discussed in Ref. 6 (see Section 6.1).

Williams and Janata[11] found that the stoichiometry of the barium sulfate precipitation depends on the concentration of ethanol when an ethanol–water mixture is used as a solvent. They suggest that solvents with less than 50% ethanol should not be used in order to avoid the adsorption of soluble sulfate on freshly precipitated barium sulfate (see Section 6.1). This effect is supposed to be a possible explanation of negative errors observed in the earlier titrations.[2,4] The sulfur content of sulfuric acid, sodium sulfate, sulfonal, and benzylthiuronium chloride was determined by sulfate titration in 70% ethanolic solution after preliminary decomposition of the organic sulfur compounds (oxygen flask combustion or fusion with sodium peroxide). The relative standard deviations are 1 to 3% (exception: sulfonal after oxygen flask combustion 7%). Interferences of various inorganic compounds were also investigated.

6.2.10 CATALYTIC THERMOMETRIC TITRATIONS

The use of catalytic end-point indication in thermometric precipitation titrations is described by Weisz, Kiss, and Klockow.[46] The titration of Ag^+, Hg^{2+}, and Pd^{2+} with iodide standard solutions (10^{-2} to 10^{-4} N KI) at the microgram level is possible when the iodide-catalyzed reaction between cerium(IV) and arsenic(III) serves as an indicator reaction.[47] Table 6.4 gives the mean values and standard deviations from nine determinations for each titration.

The determination of Cl^-, Br^-, I^-, and SCN^- is possible in the same concentration range by adding a 100% excess of standard $AgNO_3$ solution and back-titrating with KI; that of Cl^-, Br^-, I^-, SCN^-, $[Fe(CN)_6]^{4-}$, S^{2-}, and CN^- by the addition of an excess of standard $Hg(OOCCH_3)_2$ solution and back-titrating; and that of I^-, SCN^-, and CN^- also by the addition of standard $PdCl_2$ solution and back-titrating, using the same end-point indication of the back titration as before.[46]

TABLE 6.4. Titration of Ag^+, Hg^{2+}, and Pd^{2+} with KI (10^{-2} to 10^{-4} N) and Catalytic End-Point Indication[46]

| Ag^+ | | | Hg^{2+} | | | Pd^{2+} | | |
| μg | | | μg | | | μg | | |
Taken	Found	σ	Taken	Found	σ	Taken	Found	σ
431.5	431.1	1.21	318.0	318.8	0.50	280.4	281.4	0.84
420.7	419.1	0.67	180.5	181.3	0.18	160.3	159.6	0.67
4.10	4.06	0.02	4.04	4.02	0.02	4.32	4.30	0.02
1.93	1.91	0.01	1.34	1.34	0.01	1.32	1.31	0.01

REFERENCES

1. P. Dutoit and E. Grobet, *J. Chim. Phys.*, **19**, 324 (1922).
2. P. M. Dean and O. O. Watts, *J. Am. Chem. Soc.*, **46**, 855 (1924).
3. P. M. Dean and E. Newcomer, *J. Am. Chem. Soc.*, **47**, 64 (1925).
4. C. Mayr and J. Fisch, *Z. Anal. Chem.*, **76**, 418 (1929).
5. B. C. Tyson, Jr., W. H. McCurdy, Jr., and C. E. Bricker, *Anal. Chem.*, **33**, 1640 (1961).
6. J. Barthel, N. G. Schmahl, and K. Lenz, *Z. Anal. Chem.*, **233**, 328 (1968).
7. Landolt-Börnstein, *Zahlenwerte und Funktionen*, Vol. 1, Springer, Berlin–Göttingen–Heidelberg, 1955, Part 4, p. 541.
8. M. J. Sienko and R. A. Plane, *Physikalische anorganische Chemie*, S. Hirzel, Stuttgart, 1965.
9. G. J. Ewing and C. J. Mazac, *Anal. Chem.*, **38**, 1575 (1966).
10. F. Seel, *Grundlagen der analytischen Chemie*, 5th ed., Verlag Chemie, Weinheim, 1970.
11. M. B. Williams and J. Janata, *Talanta*, **17**, 548 (1970).
12. J. J. Christensen, R. M. Izatt, L. D. Hansen, and J. A. Partridge, *J. Phys. Chem.*, **70**, 2003 (1966).
13. J. C. Wasilewski, P. T. -S. Pei, and J. Jordan, *Anal. Chem.*, **36**, 2131 (1964).
14. H. W. Linde, L. B. Rogers, and D. N. Hume, *Anal. Chem.*, **25**, 404 (1953).
15. P. T. Priestley, *Analyst*, **88**, 194 (1963).
16. H. J. V. Tyrrell and A. E. Beezer, *Thermometric Titrimetry*, Chapman and Hall, London, 1968.
17. T. Takeuchi and M. Yamazaki, *Kogyo Kagaku Zasshi*, **72**, 1263 (1969).
18. C. E. Johansson, *Talanta*, **17**, 739 (1970).
19. P. Deschamps, A. Deburck, and Y. Bonnaire, *Anal. Chim. Acta*, **40**, 259 (1968).
20. W. L. Everson and E. M. Ramirez, *Anal. Chem.*, **39**, 1771 (1967).
21. J. L. Rasmussen and T. Nielsen, *Acta Chem. Scand.*, **17**, 1623 (1963).
22. P. Mondain-Monval and R. Pâris, *Bull. Soc. Chim. Fr.*, **5**, 1641 (1938).
23. P. Mondain-Monval and R. Pâris, *C. R. Acad. Sci. (Paris)*, **198**, 1154 (1934).
24. R. Pâris, *C. R. Acad. Sci. (Paris)*, **199**, 863 (1934).
25. I. Sajó and B. Sipos, *Tonind.-Ztg. Keram. Rundsch.*, **92**, 88 (1968).
26. I. Sajó and B. Sipos, *Radex-Rundsch.*, **1968**, 178.
27. I. Sajó, *Épitöanyag*, **21**, 249 (1969).
28. I. Sajó, *Magy. Kém. Foly.*, **75**, 1 (1969).
29. W. L. Everson, *Anal. Chem.*, **43**, 201 (1971).
30. J. Rondeau, M. Legrand, and R. A. Pâris, *C. R. Acad. Sci. (Paris)*, **263c**, 579 (1966).
31. M. P. Ben-Yair, *Trans. Chalmers Univ. Technol., Gothenburg*, **236**, 3 (1961).
32. P. W. Carr, *Anal. Chem.*, **43**, 756 (1971).
33. P. W. Carr, *Thermochim. Acta*, **2**, 505 (1971).
34. L. S. Bark and J. K. Grime, *Analyst*, **97**, 911 (1972).
35. P. W. Carr and J. Jordan, *Anal. Chem.*, **44**, 1278 (1972).
36. G. L. Clark and R. S. Sprague, *Anal. Chem.*, **24**, 688 (1952).
37. K. K. Chatterji, *J. Indian Chem. Soc.*, **32**, 366 (1955).

38. J. Jordan and E. J. Billingham, Jr., *Anal. Chem.*, **33**, 120 (1961).
39. J. Peisach and F. Brescia, *J. Am. Chem. Soc.*, **76**, 5946 (1954).
40. I. Sajó and B. Sipos, *Z. Anal. Chem.*, **222**, 23 (1966).
41. I. Sajó and B. Sipos, *Zem.-Kalk-Gips*, **21**, 32 (1968).
42. I. Sajó and B. Sipos, *Bull. Soc. Fr. Céram.*, **85**, 3 (1969).
43. I. Sajó and B. Sipos, *Mikrochim. Acta*, **1967**, 248.
44. I. Sajó and B. Sipos, *Talanta*, **14**, 203 (1967).
45. H. Perchec and B. Gilot, *Bull. Soc. Chim. Fr.*, **1964**, 619.
46. H. Weisz, T. Kiss, and D. Klockow, *Z. Anal. Chem.*, **247**, 248 (1969).
47. H. Weisz and U. Muschelknautz, *Z. Anal. Chem.*, **215**, 17 (1966).

CHAPTER

7

THERMOMETRIC REDOX TITRATIONS

The use of the thermometric method in redox titrations is more than just another type of application to round off a universal method. It is true that redox titrations are the special field for electrochemical methods, but some points can be mentioned that may speak for the application of the thermometric method in this domain also: Most redox reactions have a considerable heat of reaction (see Table 7.1). This means that the thermometric method is still dependable even at low concentrations, where other methods become unreliable. Back titration is avoided in many cases by use of thermometric methods, either by extrapolation of the titration end point from adjacent branches of the thermogram or by DIE technique. It is independent of interfering electrode reactions.

The previous remarks can be illustrated by the following example: The titration of Fe^{2+} with $Cr_2O_7^{2-}$ ($\Delta H^0 = -24$ kcal val^{-1}) with a sample of 0.001 M Fe^{2+} solution in 1 N H_2SO_4 is possible with an accuracy[1] of 0.3%. A 0.0001 M solution still has an accuracy better than 3%.

The thermograms of redox reactions do not display any particular traits. Figure 7.1 shows the thermogram of a titration with complete reaction, and Fig. 7.2 that of an incomplete reaction. The end point of Fig. 7.1 is sharp; that of Fig. 7.2 can be extrapolated from the neighboring branches of the curve. Table 7.1 contains a survey of the application of the thermometric method to typical redox titrations.[1] The solutions were prepared in such concentrations that a 200-ml sample of A would require 10 ml of titrant B. To facilitate comparison, all the values obtained by thermometric titration have been recalculated for 10.00 ml, and the accuracy of the titrations as well as of the comparison titrations refer to this amount. To simplify evaluation of the ΔH_R values, the sample and titrant were dissolved in the same solvent so as to eliminate solvent changes on the addition of the titrant B that might produce additional heat effects. The values for C_0 and c_B were calculated by customary procedures (see Section 3.1.2).

7.1 POTASSIUM DICHROMATE TITRATIONS

Potassium dichromate as a titrant (Table 7.1, 1a and b) has been used before in the thermometric redox titration Fe^{2+}/Fe^{3+}, although with a

TABLE 7.1. Thermometric Redox Titrations.[1]

Sample	Titrant	Solution	End Point	Comparison Value	$-\Delta H_R$, 25°C kcal val^{-1}	Comparison Titration
1a $K_2Cr_2O_7$ 0.003 M	$(NH_4)_2Fe(SO_4)_2$ 0.2–1.0 N	H_2SO_4	10.00 ± 0.03	9.96 ± 0.05	24.0 ± 0.5	With diphenylamine as indicator
b $(NH_4)_2Fe(SO_4)_2$ 0.002 M	$K_2Cr_2O_7$	0.5 N H_2SO_4	10.00 ± 0.03	10.02 ± 0.05	24.1 ± 0.5	1. With diphenylamine as indicator 2. Conductometrically[2]
c $TiCl_3$ 0.01 M	$K_2Cr_2O_7$	1 N HCl	10.0 ± 0.1	—	33 ± 1	—
d CuCl 0.025 M	$K_2Cr_2O_7$	1 N HCl	10.0 ± 0.1	—	33.5 ± 1	—
2a $(NH_4)_2Fe(SO_4)_2$ 0.002 M	$KMnO_4$	0.5–1.0 N H_2SO_4	10.00 ± 0.03	10.01 ± 0.05	23.3 ± 0.5	Self-indication of MnO_4
b $SnSO_4$ 0.005 M	$KMnO_4$	1 N H_2SO_4	10.0 ± 0.1	—	36 ± 2	
3a $Ce(SO_4)_2$ 0.01 M	$(NH_4)_2Fe(SO_4)_2$	0.2 N H_2SO_4	10.00 ± 0.03	9.98 ± 0.05	24.4 ± 0.5	Ferroin as (25°C) indicator

b	Ce(SO$_4$)$_2$ 0.01 M	(NH$_4$)$_2$Fe(SO$_4$)$_2$ 0.5 N H$_2$SO$_4$	10.00 ± 0.03	—	24.3 ± 0.5 (20°C) 23.0 ± 0.5 (25°C)
c	Ce(SO$_4$)$_2$ 0.01 M (titration; cf. Fig. 6.1)	(NH$_4$)$_2$Fe(SO$_4$)$_2$ 1.0 N H$_2$SO$_4$	10.00 ± 0.03	—	23.3 ± 0.5 (20°C) 22.5 ± 0.5 (25°C)
d	SnSO$_4$ 0.005 M	Ce(SO$_4$)$_2$ 1.0 N H$_2$SO$_4$	10.0 ± 0.1	—	—
4	H$_3$AsO$_3$ 0.025 N	I$_2$ Phosphate buffer pH 6.6	10.00 ± 0.02	10.02 ± 0.05	— Starch as indicator
5	Na$_2$S$_2$O$_3$ 0.01 M	I$_2$ Neutral	10.00 ± 0.03	10.00 ± 0.03	— Starch as indicator
6	(NH$_4$)$_2$S$_2$O$_8$ 0.0125 N	(NH$_4$)$_2$Fe(SO$_4$)$_2$ 0.1–1.0 N H$_2$SO$_4$	10.00 ± 0.5	9.98 ± 0.05	34 ± 2 Back titration of excess Fe^{2+} with MnO$_4^-$
	(NH$_4$)$_2$S$_2$O$_8$ 0.0125 N (titration; cf. Fig. 6.2)	(NH$_4$)$_2$Fe(SO$_4$)$_2$ 0.1–0.5 M H$_3$PO$_4$	10.00 ± 0.05	10.00 ± 0.05	

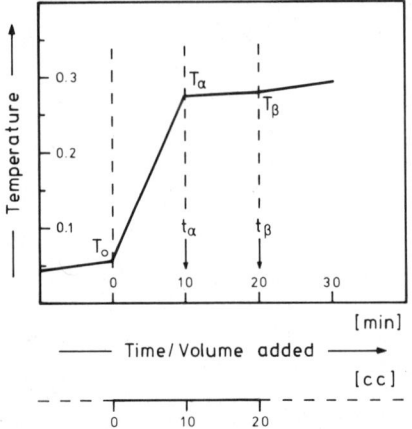
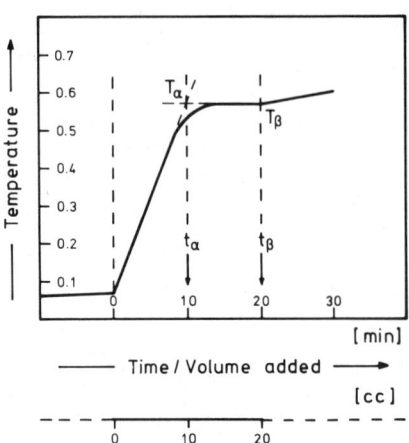

Fig. 7.1. Thermogram of the titration of $Ce(SO_4)_2$ with $(NH_4)_2Fe(SO_4)_2$ ($V_0 = 200$ ml, $a = 0.01\ M$, $b = 0.2\ M$) in $0.5\ M\ H_2SO_4$ as solvent.[1]

Fig. 7.2. Thermogram of the titration of $(NH_4)_2S_2O_8$ with $(NH_4)_2Fe(SO_4)_2$ ($V_0 = 200$ ml, $a = 0.0125\ N$, $b = 0.25\ N$) in $0.5\ M\ H_2SO_4$ as solvent.[1]

higher concentration of sample;[3] it can also be used successfully for the titration of Ti^{3+} and Cu^+ (Table 7.1, 1c and d).

Hydroquinone, even in $M/4000$ concentration, can be determined in sulfuric acid solution of dichromate by titrating the excess dichromate with $2\ M$ sodium thiosulfate.[4] The titration method used for the determination of uranium(IV), which was described by Kolthoff and Lingane,[5] was developed into a thermometric method by Miller and Thomason;[6] with a 5-mg uranium(IV) sample the standard deviation is 1%.

7.2 TITRATIONS WITH PERMANGANATE

Titrations with permanganate of iron(II) yield accurate results (Table 7.1.2a; see also Refs. 7 and 8). The titrations of oxalic acid, hydrogen peroxide, ferrous sulfate, and potassium ferrocyanide with permanganate, conducted by Mayr and Fisch,[9] were the first redox titrations with the thermometric method. The titration of sulfide with permanganate was performed by Sajó and Sipos[10] in order to determine the sulfur content of furnace slags; the oxidation of Mn(II) to Mn(IV) with $KMnO_4$ is a method widely used by these authors in the course of industrial analysis.[10-15] Further examples are the titration of Cr(III),[16] Sn(II) (see Table 7.1 2b),[1] KI^4 and H_2SeO_3.[17]

7.3 TITRATIONS WITH METAL IONS

Titrations of Ce^{4+}/Ce^{3+} are listed in Table 7.1. For comparison Jordan[18] states the accuracy of Fe^{2+} ($4 \cdot 10^{-4}$ M) titration with Ce^{4+} to be 2% with a heat effect of $\Delta H_R = -24.2$ kcal mol^{-1}. Highly accurate results were obtained by Brown, Issa, and Sinclair[19] for this titration, and Guillot[20] used it as a test reaction for the automatization of the DIE. The dependence of the reaction heat (Table 7.1) on the acidity of the sample can be explained through the different heats of formation of the various complexes of cerium.[1,21,22] The titration of Sn^{2+} with Ce^{4+} is included in Table 7.1 as a further example; the thermogram of this reaction suggests the existence of a possible intermediate step in the titration.[1] Many cerium(IV) titrations in sulfuric acid solution, including phenol oxidation, were published by Priestley.[4] Table 7.2 contains information about these titrations.

TABLE 7.2. Titrations with $M/4$ Cerium(IV) Sulfate in Sulfuric Acid Solution[4]

Sample	Molarity of Sample	Replicates	Standard Deviation %
KI	$M/120$	8	1.5
Na_2SO_3	$M/240$	13	0.7
$Na_2S_2O_3$	$M/240$	8	0.4
Hydroquinone	$M/600$	8	0.4
Metol	$M/240$	7	0.4
Phenidone	$M/600$	11	1.8
Resorcinol	$M/480$	8	0.8

Alexander, Mash, and McAuley reported titration with Ce(IV) of α-mercaptocarboxylic acids,[23] and thiourea and its N-substituted derivatives.[24] The titration of ferrous ethylenediammonium sulfate with cerium-(IV) sulfate in 3 M H_2SO_4 gives 99.8 ± 0.7% of the theoretical value, and the reaction heat[25] is stated to be 40 kcal mol^{-1}. The same publication reports the titration of H_2O_2 and $K_4Fe(CN)_6 \cdot 3H_2O$ with cerium(IV) sulfate.

Among the various methods that have been proposed for the determination of nitro and nitroso compounds,[26] the use of metal ions to reduce the functional group is often employed. Bark and Bate[27] developed a thermometric method in which the determination is conducted by reduction with a known and excess amount of titanium(III) chloride solution and subsequent titration of the excess with an iron(III) solution. The shape of the thermogram depends on the solvent that is used to dissolve the nitro and nitroso compound, and on the variation in the acidity of the system. Most advantageous are those solvents (such as ethanol, acetone, and glacial acetic acid) that are completely miscible with the aqueous solutions subsequently used. The titration is carried out in oxygen-free atmosphere. Table 7.3 contains examples.[27]

TABLE 7.3. Results for the Determination of Nitro and Nitroso Compounds[27]

Compound	Amount mg	
	Taken	Found
Nitrobenzene	6.96	6.97
	8.82	8.81
1.3-Dinitrobenzene	5.18	5.13
2.4-Dinitrobenzoic acid	4.68	4.71
4-Nitrobenzoic acid	8.69	8.70
4-Nitrophenol	8.46	8.49
3-Nitrophthalic acid	12.56	12.47
4-Nitroaniline	8.50	8.46
8-Nitroquinoline	10.22	10.18
4-Nitroacetanilide	10.03	10.00
1-Nitroso-2-naphthol	11.41	11.35
2-Nitroso-1-naphthol	13.24	13.20
4-Nitroso-N,N'-diethylaniline	14.23	14.14
N-Nitrosodiphenylamine	9.47	9.50

The titration of $S_2O_8^{2-}$ with Fe^{2+} is included in Table 7.1.6 as an example of an incomplete titration. The thermogram of the titration is reproduced in Fig. 7.2; the end point can be constructed from adjoining branches of the curve. The reverse reaction is used by Sajó and Sipos[10] for determining the iron(II) content of slags. The chromium content of chromium-plating solutions[16] was determined as Cr(VI) with Fe(II).

7.4 HYDROGEN PEROXIDE TITRATIONS

Hydrogen peroxide was used as a titrant by Rivenq[28,29] in the formation of peroxides of vanadium, chromium, molybdenum, tungsten, titanium, and uranium. The thermograms of some titration reactions, for example,

$$V_2O_5 + 2H_2O_2 \rightleftharpoons V_2O_7 + 2H_2O$$

show a sharp break indicating the peroxide compound [also Cr(VI), Ti(IV), and Mo(VI)], whereas others (e.g., $UO_3 + H_2O_2$) do not. The reaction of vanadium(V) with hydrogen peroxide was also used by Sajó[15] in the analysis of silicates; that of titanium(IV)[10] and iron(II)[30] in the analysis of furnace slags; and that of molybdenum(VI) for the determination of molybdenum in alloys.[31] The determination of sulfite and bisulfite with hydrogen peroxide was performed by Štráfelda and Hájková.[32]

7.5 TITRATIONS WITH ARSENIOUS ACID, IODINE, HYPOCHLORITE, AND FURTHER AGENTS

Arsenious acid is used as sample in the titration listed in Table 7.1.4. The titration of arsenite can be performed with iodine in a phosphate buffer (pH = 6.6) without difficulty. The thermometric method cannot be used with bicarbonate buffer, whereas the classical method produces adequate results in this case also;[1] but even if this method is used, precision titration should not be made in a carbonate buffer.[33] According to Mayr and Fisch[9] arsenious acid can be determined accurately by titration with potassium bromate at a definite acidity. The active chlorine of a solution of hypochlorite can be titrated with arsenious acid that has been standardized iodometrically.

Several authors[1,4] have discussed the titration of thiosulfate with iodine (Table 7.1.5). Priestley[4] reports that it is possible to determine $(2\,M)$ I_2 in the presence of KI $(M/60)$, $Ce(SO_4)_2$ in H_2SO_4 $(M/30)$. Na_2SO_3 in KI_3 $(M/1000)$, and $NaHSO_3$ in KI_3 $(M/1000)$ with the help of thiosulfate. The two last-mentioned titrations are achieved by determining the excess of I_2. Olefinic unsaturation can be determined by iodine titration.[34] ICl is added in excess to the olefin sample and the unused ICl reacted with KI $(ICl + I^- \rightarrow I_2 + Cl^-)$. The I_2 set free during the reaction is titrated thermometrically.

Billingham and Reed[35] successfully determined copper iodometrically.

In the present context we must also mention the determination of elementary chlorine with potassium iodide,[36] the hypochlorite determination in fabric-bleaching solutions,[37] the titration of iodide and thiosulfate with hypochlorite,[19] and a comprehensive study by Schäfer and Wilde of the titration of sulfonamides with hypochlorite.[38]

Ascorbic acid was used as reduction agent in the titration[10] of chromium(VI); formaldehyde in the reduction of manganese(VII).

The peroxodisulfate titration of Fe(II) (see Section 7.3) has been elaborated by the authors to determine the total iron content of slags, silicates, magnesites, dolomites, clays, cements, and clinkers.[10-15] For this purpose the Fe(III) content of the samples is at first reduced to Fe(II) by convenient agents (polysulfide, stannous chloride); then Fe(II) is titrated as outlined earlier.

REFERENCES

1. J. Barthel and N. G. Schmahl, *Z. Anal. Chem.*, **207**, 81 (1965).
2. W. Walisch and J. Barthel, *Z. Phys. Chem.* (New Series), **39**, 235 (1963).
3. G. W. Ewing, *Instrumental Methods of Chemical Analysis*, McGraw-Hill, New York, 1954.
4. P. T. Priestley, *Analyst*, **88**, 194 (1963).
5. I. M. Kolthoff and J. J. Lingane, *J. Am. Chem. Soc.*, **55**, 1871 (1933).
6. F. J. Miller and P. F. Thomason, *Anal. Chim. Acta*, **21**, 112 (1959).
7. R. H. Müller, *Ind. Eng. Chem.*, Anal. Ed., **13**, 667 (1941).
8. P. T. Priestley, W. S. Sebborn, and R. F. W. Selman, *Analyst*, **88**, 797 (1963).
9. C. Mayr and J. Fisch, *Z. Anal. Chem.*, **76**, 418 (1929).
10. I. Sajó and B. Sipos, *Z. Anal. Chem.*, **222**, 23 (1966).
11. I. Sajó, *Kém. Közl.* (Budapest), **26**, 119 (1966).
12. I. Sajó and B. Sipos, *Zem.-Kalk-Gips*, **21**, 32 (1968).
13. I. Sajó and B. Sipos, *Bull. Soc. Fr. Céram.*, **85**, 3 (1969).
14. I. Sajó and B. Sipos, *Radex-Rundsch.*, **1968**, 178.
15. I. Sajó, *Épitöanyag*, **21**, 249 (1969).
16. I. Sajó and B. Sipos, *Talanta*, **14**, 203 (1967).
17. M. J. Dastoor and B. C. Haldar, *Indian J. Chem.*, **5**, 335 (1967).
18. J. Jordan, *Chimia*, **17**, 101 (1963).
19. M. W. Brown, K. Issa, and A. G. Sinclair, *Analyst*, **94**, 234 (1969).
20. P. Guillot, *Anal. Chim. Acta*, **50**, 499 (1970).
21. T. J. Hardwick and E. Robertson, *Can. J. Chem.*, **29**, 828 (1951).
22. E. G. Jones and F. G. Soper, *J. Chem. Soc.*, **1935**, 802.
23. W. A. Alexander, C. J. Mash, and A. McAuley, *Talanta*, **16**, 535 (1969).
24. W. A. Alexander, C. J. Mash, and A. McAuley, *Analyst*, **95**, 657 (1970).
25. B. C. Tyson, Jr., W. H. McCurdy, Jr., and C. E. Bricker, *Anal. Chem.*, **33**, 1640 (1961).
26. M. R. F. Ashworth, *Titrimetric Organic Analysis*, Interscience, New York, Part I (1964); Part II (1965).
27. L. S. Bark and P. Bate, *Analyst*, **98**, 103 (1973).
28. F. Rivenq, *Bull. Soc. Chim. Fr.*, **12**, 283 (1945).
29. F. Rivenq, *Bull. Soc. Chim. Fr.*, **12**, 289 (1945).
30. I. Sajó, *Kohász. Lapok*, **7**, 287 (1957).
31. I. Sajó, W. Foerster, H. Rüdiger, and A. Sipos, *Neue Hütte*, **12**, 500 (1967).
32. F. Štráfelda and M. Hájková, *Collect. Czech. Chem. Commun.*, **33**, 4389 (1968).
33. E. W. Washburn, *J. Am. Chem. Soc.*, **30**, 31 (1908).
34. J. Jordan, P. T.-S. Pei, and R. C. Buchta, Jr., *Abstr. 149th Nat. Meeting, Am. Chem. Soc.*, Detroit, 1965, p. 30B.
35. E. J. Billingham, Jr., and A. H. Reed, *Anal. Chem.*, **36**, 1148 (1964).
36. P. Marik-Korda and L. Erdey, *Talanta*, **17**, 1215 (1970).
37. F. Štráfelda and J. Kroftová, *Collect. Czech. Chem. Commun.*, **33**, 4171 (1968).
38. H. Schäfer and E. Wilde, *Z. Anal. Chem.*, **130**, 396 (1949/50).

CHAPTER

8

FORMATION OF COMPOUNDS AND OF COMPLEXES.
PROBLEMS OF COMPLEXOMETRIC TITRATION

The results contained in the publications in this field are conveniently classified into the following three groups:
1. Determination of the composition of a complex, sometimes with an attempt also to determine the heat of complex formation.
2. Determination of thermodynamic data of complex formation with the help of an entropy titration.
3. Complexometric determination of the end point, usually combined with an approximate determination of the thermodynamic data of the complex compound.

There are no sharp borderlines between these three aims. The methods of investigating true complex compounds and other chemical compounds according to the first two points mentioned are similar; no sharp distinction will thus be made here.

8.1 ATTEMPTS TO DETERMINE THE COMPOSITION OF COMPLEX COMPOUNDS

A few investigations into this problem have already been mentioned in the preceding chapters, for example, Refs. 1–4, and 55 discussed in the chapters on precipitation and Lewis acid titration. It is the aim of such experiments to determine the composition of the compound from the amount of titrant used up to a break in the thermogram. It can then be decided whether this indication can be used for the development of a titration method. The problems encountered in this method are variations of those of the classical thermometric titration methods.

The thermometric method was used in many investigations on multistep reactions of complex formation in aqueous solution, establishing the individual steps of the complex formation but without taking into account equilibria depending on concentration and the thermal powers of the reaction depending on it. The technology of these early titrations did not enable heats of dilution, heats of mixing, and heat exchange with the environment to be accounted for, thus causing additional sources of error.

According to what has been said in Chapter 2, sharp breaks in the titration curve will be obtained only for compounds with a large constant of complex formation (stable, undissociated compounds). It is sometimes unreliable to construct the end points for partly dissociated compounds by extrapolating from the adjoining parts of the curve or by constructing tangents for a more or less regularly curved line and hoping that the point of intersection of such linear sections will indicate compounds. In this connection mention must be made of the question, discussed in Chapter 4, of whether it is possible to titrate two acids simultaneously, and the problem of indication of the partial neutralization of polyvalent acids. All these problems have the same theoretical background.

Haldar investigated the basic sulfates of bivalent metals and concluded from thermograms of the type $\Delta T = f(v)$ (the type of Fig. 2.1) that the following compounds existed: $CdSO_4 \cdot 3CdO$ and $CdSO_4 \cdot CdO$;[5] $CuSO_4 \cdot 3CuO$ and $CuSO_4 \cdot CuO$, and $NaCu(OH)_3$ in an excess of base;[6] $ZnSO_4 \cdot 3Zn(OH)_2$;[7] $BeSO_4 \cdot BeO$;[8] and $NiSO_4 \cdot 3NiO$.[9] The same author suggests on the basis of the thermogram of the titration of CdI_2 with KI that complex ions CdI_4^{2-} and CdI_5^{3-} occur in aqueous solution.[10] The titration apparatus used[5] corresponds more or less to that of Dutoit and Grobet (see Chapter 2). In a further study on the formation of complex ions of beryllium and pyrophosphate ions, the author infers the existence of monomeric $[Be(P_2O_7)]^{2-}$ and $[Be(P_2O_7)_2]^{6-}$,[11] by using a combination of methods, conductivity measurements, thermometric titrations, pH measurements, cryoscopic procedures, and determinations of transport numbers. Thermometric and conductometric titration prove the existence of $K_4UO_2(CO_3)_3$, $K_2UO_2(CO_3)_2$, and the corresponding sodium compounds.[12]

Purkayastha reports the existence of the four fluoroberyllate ions $[BeF_3]^-$, $[BeF_4]^{2-}$, $[BeF_5]^{3-}$, and $[BeF_6]^{4-}$,[13] and the formation of the complex cation $[Pb(C_2H_3O_2)]^+$ when lead salts are treated with sodium acetate;[14] he employed conductometric and thermometric methods.

Banerjee[15] infers the existence of the ions $Fe(HPO_4)^+$ and $Fe(HPO_4)_2^-$ from the titration thermogram of $FeCl_3$ with H_3PO_4 and from conductivity measurements.

The composition of complex ferro- and ferricyanides was studied by Bhattacharya et al. The breaks in the corresponding titration curves lead to the conclusion that the following compounds exist: $K_2CuFe(CN)_6$ and $K_2Cu_3[Fe(CN)_6]_2$,[16] $K_2CdFe(CN)_6$,[17] $KCd_{10}[Fe(CN)_6]_7$,[18] $Co_3[Fe(CN)_6]_2$,[19] $Cu_3[Fe(CN)_6]_2$,[20] $KFe[Fe(CN)_6]$ and $Fe_4[Fe(CN)_6]_3$,[21] $KFe[Fe(CN)_6]$ and

$Fe_3[Fe(CN)_6]$,[22] $Fe[Fe(CN)_6]$,[23] $Zn_3[Fe(CN)_6]_2$,[24] $Ni_3[Fe(CN)_6]_2$ and $KNi_4[Fe(CN)_6]_3$,[25] and $FeK_2[Fe(CN)_6]$ and $Fe_2[Fe(CN)_6]$.[26] The titrations were conducted with the apparatus mentioned above[5] in water or in water–alcohol mixtures. Studies of the complex compositions were also made with the help of different physicochemical methods in addition to the thermometric titration. The compounds $Ag_3Fe(CN)_6$,[27] $KCe[Fe(CN)_6]$,[28] and $K_2UO_2Fe(CN)_6 \cdot xUO_2(NO_3)_2$ have also been reported.[29]

Gaur and Bhadraver[30] found silver thiosulfate complexes of composition $Na_3Ag(S_2O_3)_2$, $NaAgS_2O_3$, $Na_5Ag_3(S_2O_3)_4$, and $Ag_2S_2O_3$ in the thermograms of their titration of $AgNO_3$ with $Na_2S_2O_3$, whereas the reverse titration produced only $NaAgS_2O_3$ and $Ag_2S_2O_3$.

Thermal decomposition, conductivity titrations, and thermometric titrations of the nitrates of La, Sm, Pr, Nd with NaOH show the existence of basic nitrates of the type $XONO_3$ and $X_2O_3 \cdot 2XONO_3$.[31] Starting from the nitrates of Y, Yb, and Gd, the soluble basic nitrate $3X_2O_3 \cdot 4N_2O_5$ is obtained; in addition, gadolinium yields the insoluble salt $GdONO_3$.[32]

According to Vartak and Kabadi[33] the titration of KNO_2 with $Pb(NO_3)_2$ indicates the compounds $4KNO_2 \cdot Pb(NO_3)_2$, $2KNO_2 \cdot Pb(NO_3)_2$, and $KNO_2 \cdot Pb(NO_3)_2$; the titration[34] of $NaNO_2$ with $Pb(NO_3)_2$ gives $NaNO_2 \cdot Pb(NO_3)_2$. There is no indication of complex formation in the titration of $Pb(NO_3)_2$ with $LiNO_3$ and $NaNO_3$; titrations with K, Rb, and NH_4 nitrate[35] indicate $4RNO_3 \cdot Pb(NO_3)_2$, $2RNO_3 \cdot Pb(NO_3)_2$, and $RNO_3 \cdot Pb(NO_3)_2$.

The titration of potassium oxalate with thorium nitrate produces several thorooxalates; the existence of $K_4[Th(C_2O_4)_4]$ and $K_2[Th(C_2O_4)_3]$ is confirmed by both thermometric and cryoscopic titration.[36]

The thermometric titration curves of alkali pyrophosphate solutions with nickel and cobalt salt solutions[37] show the formation of two complex ions ($P_2O_7^{4-}$: metal = 2 : 1 and 1 : 1). The titration of phenol with $FeCl_3$ yields the compound $Fe[Fe(OPh)_6]$.[38]

Chatterji[39–41] pointed out the fact that the extrapolation of a number of nonexistent breaks of the curve in the interpretation of thermograms of the kind described can be circumvented if the volume is corrected according to (2.28) or (2.29), respectively. If Purakayasta's statements about the existence of four fluoroberyllates[13] are revised by taking into account the corrected volumes, only one compound of the composition $[BeF_4]^{2-}$ remains; the other breaks constructed from the curve are deceptions.[42] In a similar way, Haldar's interpretation of the complex ions in the thermogram of the titration of CdI_2 with KI cannot be upheld in the light of volume correction. Chatterji himself finds compounds in the molar ratios of 1 : 6, 1 : 3, 4 : 9, and 2 : 3 in a titration[43] of Cu^{2+} with $S_2O_3^{2-}$.

8.2 DETERMINATION OF THERMODYNAMIC DATA OF A COMPLEX-FORMATION REACTION

8.2.1 THE METHOD OF CONTINUOUS VARIATION

Job's method[44,45] of continuous variation consists in measuring an additive physical quality X in a series of mixtures of the reactants A (concentration of solution c) and B (concentration of solution $p \cdot c$) that react to form the complex A_xB_y. From the deviation ΔX from additivity, the composition of the complex can be determined.

The reaction of complex formation

$$x\text{A} + y\text{B} \rightleftarrows A_xB_y \tag{8.1}$$

with the dissociation constant

$$K_d = \frac{[\text{A}]^x[\text{B}]^y}{[A_xB_y]} \tag{8.2}$$

implies the following relationship of the concentrations

$$[\text{A}] + x[A_xB_y] = c(1-z) \tag{8.3}$$

$$[\text{B}] + y[A_xB_y] = cz \tag{8.4}$$

for a mixture consisting of $(1-z)$ volumes of A and z volumes of B. A graph of $\Delta X = f(z)$ shows the maximum complex concentration at the point of maximum deviation and thus furnishes the composition of the complex. Including the condition $d[A_xB_y]/dz = 0$, we get

$$\frac{cx + y^{-1}p^{y-1}[(px+y)z_0 - y]^{x+y}}{x^{y-1}y^{x-1}(p-1)^{x+y-1}} = K_d[y - (x+y)z_0] \tag{8.5}$$

If $p = 1$, that is, with equimolar solutions of A and B, the complex composition can be found as $z_0 = y/(y+x)$ on the abscissa z_0 of the maximum of the curve, independent of the numerical value of the dissociation constant K_d of the complex.

If the composition of the complex is known and if $p \neq 1$, K_d can be determined from (8.5) or with the help of Hagenmuller's procedure[46] developed from Job's method. In his critical survey Asmus[47-49] pointed out the limitations of application of Job's method: It can only be used for single-step reactions and there are several sources of erroneous results if the complexes are not of the 1:1 ($z_0 = 0.5$) type.

Job et al.[50] already used heats of reaction as additive qualities together with molar extinction and conductance. Siddhanta[51] pointed out the advantage of using the molar enthalpies of the initial reactants and final products in Job's method. Deviations from additivity of the enthalpies when the reactants are mixed must be explained by the heat of reaction of the complex formation according to (8.1). Reference 51 gives a detailed description of the method, which also respects the part played by the heats of dilution in preparing the mixtures. The special advantage of the choice of reaction enthalpy as physical quality X is that not only is K_d of the complex obtained, but also, at the same time, the enthalpy of formation, $\Delta H_f(A_xB_y)$. A method of entropy titration is thus developed in principle, at least for single-step reactions.

Siddhanta and Guha[52] determined a complex of composition $[Cu(NH_3)_4]^{2+}$ with a dissociation constant $K_d = 3 \cdot 10^{-9}$ (30°C) in a series of mixtures of equimolar solutions ($p=1$) and solutions of $CuSO_4$ and NH_4OH of different molarities ($p \neq 1$). In view of the fact that the method of investigation had not been developed extensively, this result corresponds satisfactorily with the result obtained by Job's method, using extinction measurements,[45] in the same system [$K_d = 0.5 \cdot 10^{-9}$ (16°C)]. A further application of Job's thermometric method, namely, the identification of the compound CdS_2O_3, is reported by Bhadraver and Gaur.[53]

8.2.2 OTHER METHODS FOR THE APPROXIMATE DETERMINATION OF THE CONSTANT OF FORMATION K_c

Cioffi and Zenchelsky[55] have proposed a graphical method for determination of the degree of dissociation α of a compound from the thermogram of a titration with incomplete titration reaction (see Section 2.3.7). At the stoichiometrically calculated equivalence point v_α (Fig. 8.1), the degree of dissociation will be $\alpha = (T'_\alpha - T_\alpha)/(T'_\alpha - T'_0)$. The point T'_0 is the point of intersection of the extrapolated preperiod and the isologue ($v_\alpha = $ const) drawn at the stoichiometrically calculated equivalence point. The construction of the point T'_α is based, according to the authors, on a calculation of ΔH^0_R, which must be determined in a separate calorimeter experiment. The heat of mixing and the heat exchange with the environment can be taken into consideration. The construction $T'_\alpha - T'_0$ eliminates the heat caused by stirring. To derive the equilibrium constant from α, the composition of the complex must be determined first.

The particular problem discussed by the authors is the titration of $SnCl_4$ with tetrahydrofuran (THF), tetrahydropyran, 2-methyltetrahydrofuran, 2,5-dimethyltetrahydrofuran, and 4-methyltetrahydropyran. In these cases,

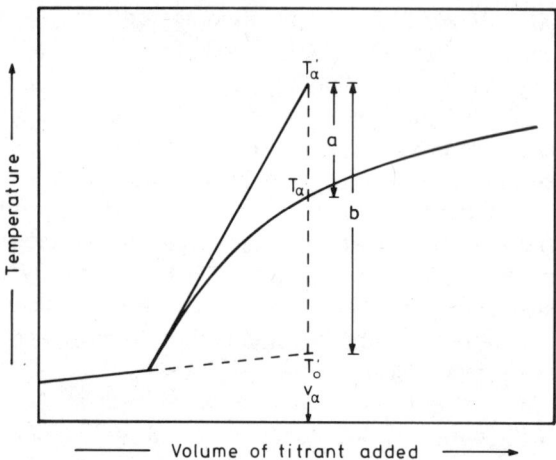

Fig. 8.1. Thermometric titration curve used to calculate K.[55] The fraction dissociated at the equivalence point, α, equals a/b.

all complexes are of the type AB_2 and the incomplete reaction is of the following type:[54,55]

$$SnCl_4 + 2THF \rightleftharpoons SnCl_4 \cdot 2THF$$

The constant of formation, calculated at the equivalence point, is

$$K_c = \frac{1-\alpha}{4\alpha^3 \cdot c^2}$$

in which c is the analytical concentration of $SnCl_4$. Because $10^{-3} < c < 2 \cdot 10^{-2}$ mol liter^{-1}, we can assume that $K_c \approx K_a$.

Another graphical method is that proposed by Dilke and Eley.[56] The method proposed by Cioffi and Zenchelsky has been further developed by Papoff and Zambonin[57] on the basis of the complete thermometric equation (2.19) (see also Section 2.3.7).

8.2.3 ENTROPY TITRATION

The exact procedures for determination of the complete thermodynamic data of complex formation are described by the term *entropy titration*, coined by Christensen et al.[58,59] Early investigations in this field are due to Izatt et al.,[60] Bjerrum et al.,[61] Sillén et al.,[62,63] and Schlyter et al.[64-67]

In Section 2.3.6 the importance of this method was discussed for the example of a single-step reaction

$$A + B \rightleftharpoons AB$$

Examples of single-step complex formation and an evaluation of their thermograms for ΔH_R^0, ΔS_R^0, and ΔG_R^0 (or K_a) can be found in the investigations of Becker and Grundmann[68] on Cu^{2+}/SO_4^{2-} and Cd^{2+}/SO_4^{2-} reactions. Other single-step reactions that have been evaluated as entropy titrations are $HSO_4^- \rightleftharpoons SO_4^{2-} + H^+$ [58], mentioned previously in our discussion of the foundations of the method; $HPO_4^{2-} \rightleftharpoons PO_4^{3-} + H^+$;[57,58,69] the reactions of several pentoses, hexoses, and their monophosphates;[70] of pyrimidines and their nucleotides;[71,72] of cysteine and mercaptoacetic acid;[73] of ephedrine and pseudophedrine;[74] and of substituted tetrazoles;[75] and the hydrolysis equilibrium[76] of $Cu(ClO_4)_2$.

When conducted in reverse also, that is, by interchanging titrand and titrant, thermometric titration can offer proof of whether there is a true single-step equilibrium. Thus a thermometric investigation of the reaction of phenol with dimethylformamide established that an excess of phenol produced a compound of the composition phenol:dimethylformamide $= 2:1$,[79] whereas according to the existing literature on this reaction,[77,78] only a 1:1 molecular compound could be formed in a single-step reaction. Probably, a second phenol molecule is attached to the second free electron pair of the carbonyl oxygen of the dimethylformamide via a somewhat weaker H bond.[79] These titrations were performed in both isooctane and toluene as solvents.

A suitable example for discussion of the different methods of evaluating a multistep reaction process is the system Ag^+/pyridine. It is a two-step equilibrium (Py = pyridine) and has been investigated repeatedly,[80–82] using every evaluation procedure known.

$$Ag^+ + Py \rightleftharpoons Ag^+Py$$

$$Ag^+Py + Py \rightleftharpoons Ag^+Py_2$$

Its data can be reproduced with sufficient accuracy (see Table 8.3), for judging different methods and instrumentation of entropy titration. Moreover, corresponding data obtained by nonthermometric methods are available.[83,84] The system was first investigated by Becker et al.[80] The following paragraph discusses the theory used for the evaluation of an N-step equilibrium, which is the basis of this and later methods.

In an N-step equilibrium

$$\left.\begin{array}{c} A + B \rightleftharpoons AB \\ AB + B \rightleftharpoons AB_2 \\ \vdots \\ AB_{N-1} + B \rightleftharpoons AB_N \end{array}\right\} \quad (8.10)$$

the individual constants of formation k_i and the overall constant of formation K_i are

$$\frac{k_i}{V} = \frac{n_{AB_i}}{n_{AB_{i-1}} \cdot n_B} \qquad (i = 1, 2, \ldots, N) \tag{8.11}$$

$$\frac{K_i}{V^i} = \frac{n_{AB_i}}{n_A \cdot n_B^i} \qquad (i = 1, 2, \ldots, N) \tag{8.12}$$

In a first experiment, a sample V_0 (liters) of solution A (a mol liter^{-1}) is used, to which the solution B (b mol liter^{-1}) is added at constant velocity v_0' (liters min^{-1}). At the time t (min), $V = V_0 + v_0't$ (liters) solution will be in the titration vessel with

$$n_A^0 = a \cdot V_0 \quad \text{and} \quad n_B^0 = bv_0't \tag{8.13}$$

moles of A and B, respectively (for meaning of symbols see Section 2.3.3).

The *tangent method* (Section 2.3.5) can easily be applied to a multistep reaction process. Using (8.12), we can write the thermal power of the entire reaction (2.10) as

$$w_R(t) = -\sum_{i=1}^{N} \Delta H_{AB_i} \frac{dn_{AB_i}}{dt} = -\sum_{i=1}^{N} K_i \Delta H_{AB_i} \frac{d}{dt}\left(n_A \frac{n_B^i}{V^i}\right) \tag{8.14}$$

where ΔH_{AB_i} is the enthalpy of reaction of the process $A + iB \rightleftharpoons AB_i$. The function $w_R(t_0)$, which is needed for the use of the tangent method, can be derived directly from (8.14) as

$$w_R(t_0) = \lim_{t \to 0} w_R(t) = -aK_1 \Delta H_{AB}\left(\frac{dn_B}{dt}\right)_{t=0} \tag{8.15}$$

Since [see (2.33b)]

$$n_B = n_B^0 - \sum_{i=1}^{N} i n_{AB_i} = n_B^0 - \sum_{i=1}^{N} i K_i n_A \left(\frac{n_B}{V}\right)^i \tag{8.16}$$

we get

$$\left(\frac{dn_B}{dt}\right)_{t=0} = \frac{bv_0'}{1 + aK_1} \tag{8.17}$$

and hence

$$w_R(t_0) = -bv_0' \Delta H_{AB} \frac{aK_1}{1 + aK_1} \tag{8.18}$$

DETERMINATION OF THERMODYNAMIC DATA 143

Equation (8.18) is identical with (2.59) for single-step reactions; that is, the initial power and hence the initial slope in the thermogram are determined alone by the rate of formation in the first step of complex formation. Equation (2.60) can, therefore, be used also. From the initial powers w_R^I and w_R^{II}, measured for two concentrations a^I and a^{II} of the sample, we can calculate k_1

$$k_1 = K_1 = \frac{w_R^I a^{II} - w_R^{II} a^I}{a^I a^{II}(w_R^{II} - w_R^I)} \tag{8.19}$$

and ΔH_{AB} by inserting (8.19) in (8.18).

The tangent method can be used to determine all the four unknown quantities K_1, K_2, ΔH_{AB} and ΔH_{AB_2} in the case of a two-step equilibrium ($N=2$). For this, the direction of titration is reversed, that is, solution B (b^* mol liter^{-1}, $n_B^0 = b^* V_0$) is used as sample and solution A (a^* mol liter^{-1}, $n_A^0 = a^* v_0' t$) as titrant.

From (8.14) it follows that

$$w_R^*(t_0) = -\sum_{i=1}^{N} \Delta H_{AB_i} K_i b^{*i} \left(\frac{dn_A}{dt}\right)_{t=0} \tag{8.20}$$

and, since [see (2.33a)]

$$n_A = n_A^0 - \sum_{i=1}^{N} n_{AB_i} = n_A^0 - \sum_{i=1}^{N} K_i n_A \left(\frac{n_B}{V}\right)^i \tag{8.21}$$

we obtain

$$\left(\frac{dn_A}{dt}\right)_{t=0} = \frac{a^* v_0'}{1 + \sum_{i=1}^{N} K_i b^{*i}}$$

and thus

$$w_R^*(t_0) = -a^* v_0' \frac{\sum_{i=1}^{N} \Delta H_{AB_i} K_i b^{*i}}{1 + \sum_{i=1}^{N} K_i b^{*i}} \tag{8.22}$$

If the titration is performed twice in the reverse direction with different concentrations of sample, b^{*I} and b^{*II}, w_R^{*I} and w_R^{*II} can be determined. For a two-step reaction, (8.22) gives

$$\Delta H_{AB_2} = \frac{(b^{*II})^2 w_R^{*II} f^I - (b^{*I})^2 w_R^{*I} f^{II}}{a^* v_0' \left[(b^{*I})^2 f^{II} - (b^{*II})^2 f^I\right]} \tag{8.23}$$

with

$$f = w_R^* + K_1 b^* (\Delta H_{AB} a^* v_0^c + w_R^*) \tag{8.24}$$

The values K_1 and ΔH_{AB}, obtained in the first experiment, are inserted in (8.24), and this equation is combined with (8.23) to obtain ΔH_{AB_2}. Finally, K_2 can be determined from (8.22).

The *section method* alone (see Section 2.3.4) does not suffice to determine all four unknown quantities of the two-step equilibrium. It is advisable to determine K_1 and ΔH_{AB} by the tangent method.

The integral heat effect of an N-step equilibrium according to (8.10) is

$$-Q_R(t) = \sum_{i=1}^{N} n_{AB_i} \Delta H_{AB_i} = n_A \sum_{i=1}^{N} K_i \Delta H_{AB_i} \left(\frac{n_B}{V}\right)^i \tag{8.25}$$

This equation can be combined with (8.16) to give

$$\sum_{i=1}^{N} K_i \left(\frac{n_B}{V}\right)^i \left[(n_B^0 - n_B)\Delta H_{AB_i} + iQ_R\right] = 0 \tag{8.26}$$

and for the two-step equilibrium ($N=2$)

$$k_2 = \frac{K_2}{K_1} = -\frac{V\left[(n_B^0 - n_B)\Delta H_{AB} + Q_R\right]}{n_B\left[(n_B^0 - n_B)\Delta H_{AB_2} + 2Q_R\right]} \tag{8.27}$$

On the other hand, (8.21) and (8.16) result in

$$\sum_{i=1}^{N} K_i \left(\frac{n_B}{V}\right)^i (n_B^0 - n_B - in_A^0) + (n_B^0 - n_B) = 0 \tag{8.28}$$

and this gives for $N=2$

$$n_B^3 K_2 + n_B^2 \left[VK_1 - K_2(n_B^0 - 2n_A^0)\right] + n_B\left[V^2 - VK_1(n_B^0 - n_A^0)\right] - n_B^0 V^2 = 0 \tag{8.29}$$

By the use of (8.27) and (8.29) the section procedure thus leads to a method of iteration for the determination of K_2 and ΔH_{AB_2} when K_1 and ΔH_{AB} are known: an approximative value of K_2 is inserted in (8.29) as initial value $^{(0)}K_2$, and $^{(0)}n_B$ values are determined for two suitable points of time during the titration period. Using the result from this, (8.27) is solved for K_2 and ΔH_{AB_2}, and K_2 is used as $^{(1)}K_2$ in (8.29) for the next iteration.

DETERMINATION OF THERMODYNAMIC DATA

Table 8.1 contains experimental data from the evaluation of a two-step equilibrium using the tangent method.[80]

TABLE 8.1. Initial Power of the Thermometric Titration[80] of the System $Ag^+ + 2Py \rightleftarrows [AgPy_2]^+$ in Water at 25°C

Experiment Number	Sample and concentration mol liter^{-1}		Titrant and concentration mol liter^{-1}		$w(t_0) = w_R + w_v$ cal min^{-1}	$w_R(t_0)$ cal min^{-1}
1	AgNO$_3$	0.015	Py	0.6089	4.47 (4.49)	
1v	H$_2$O	—	Py	0.6089	0.280	4.19 (4.21)
2	AgNO$_3$	0.090	Py	0.6089	5.73 (5.77)	
2v	H$_2$O	—	Py	0.6089	0.280	5.45 (5.49)
3	Py	0.0482	AgNO$_3$	0.500	9.71	
3v	H$_2$O	—	AgNO$_3$	0.500	−0.287	10.00
4	Py	0.197	AgNO$_3$	0.500	10.83	
4v	H$_2$O	—	AgNO$_3$	0.500	−0.287	11.12

In parentheses: results obtained in a replicate experiment.

The results $w_R(t_0)$ of experiments 1 and 2 give, according to (8.18) and (8.19), $K_1 = 174$ (178) liter mol^{-1} and $\Delta H_{AgPy^+} = -4.77$ (−4.76) kcal mol^{-1}; experiments 3 and 4 give, based on (8.22) and (8.23), $k_2 = 89.1$ liter mol^{-1}, $K_2 = 15{,}480$ liter^2 mol^{-2} and $\Delta H_{AgPy_2^+} = -11.53$ kcal mol^{-1} (tangent method).

If the section method is used for evaluating experiment 1 according to (8.27) and (8.29), using the experimental data of table 8.2, the iteration method yields the values $k_2 = 85.6$ liter mol^{-1}, $K_2 = 14{,}870$ liter^2 mol^{-2}, and $\Delta H_{AgPy_2^+} = -12.12$ kcal mol^{-1}.

TABLE 8.2. Experimental Data[80] for Evaluation of the Thermometric Titration $Ag^+ + 2Py \rightleftarrows [AgPy_2]^+$ in Water at 25°C

t min	Q_R cal	$n_A^0 \cdot 10^3$ mol	$n_B^0 \cdot 10^3$ mol	V liter
6	23.06	3.00	7.288	0.21197
10	29.18	3.00	12.148	0.21995

The not wholly satisfactory agreement of the results can be explained by the uncertainty of the measured values.

Table 8.3 contains a survey of the results obtained with the tangent method. Also included are the results of later experiments obtained with methods of numerical evaluation[81,82] and the results of nonthermometric investigations.[83,84]

TABLE 8.3. Results of Entropy Titrations Using Various Methods for the reaction $Ag^+ + 2Py \rightleftarrows [AgPy_2]^+$ in Water at 25°C

Equilibrium	Reference	$-\Delta G^0$ kcal mol^{-1}	$-\Delta H^0$ kcal mol^{-1}	$-\Delta S^0$ cal mol^{-1} deg^{-1}
$Ag^+ + Py \rightleftarrows [AgPy]^+$	80	3.06	4.77	5.7
	81	2.73	4.83	7.0
	82	—	4.6	6.0
Nonthermometric	83, 84	2.69 2. 3	—	—
$[AgPy]^+ + Py \rightleftarrows [AgPy_2]^+$	80	2.66	6.76	13.8
	81	2.88	6.51	12.2
	82	—	6.65	13.0
Nonthermometric	83, 84	3.24 2.87	—	—
$Ag^+ + 2Py \rightleftarrows [AgPy_2]^+$	80	5.72	11.53	19.5
	81	5.61	11.34	19.2
	82	—	11.25	19.0
Nonthermometric	83, 84	5.93 5.60	—	—

Figure 8.2 shows the relative change of the number of moles of the reactants during the titration. The curves have been calculated for the conditions of experiment 1 in Table 8.1 with $K_1 = 174$ liter mol^{-1} and $K_2 = 15,000$ liter^2 mol^{-2}. The figure makes it clear which points on the curve are suitable for an application of the section method so that the relative number of moles varies as much as possible within the time interval chosen.

General numerical methods for evaluation of experimental data in a multistep equilibrium (8.10) are based on (8.21), (8.25), and (8.28), just as the section method.

Starting with a set of variables in zero approximation $^{(0)}K_i$ and $^{(0)}\Delta H_{AB_i}$, we calculate the quantities n_A and n_B with the help of (8.21) and (8.28), and then Q_R is calculated utilizing (8.25). By varying the parameters, the minimum of the function (8.30) calculated for all measured values

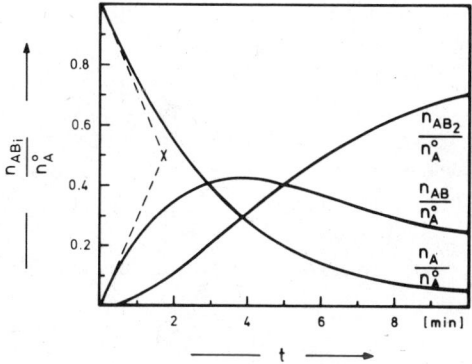

Fig. 8.2. Relative variation of the mole numbers of the reactants during the thermometric titration of silver nitrate with pyridine. (Calculated for experiment 1 in Table 8.1).[80]

$r = 1,\ldots,n$ is obtained[85,86]

$$U(K_i, \Delta H_{AB_i}) = \sum_{r=1}^{n} g_r [(Q_{R,\text{calc}})_r - (Q_{R,\text{exp}})_r]^2 \qquad (8.30)$$

The quantity g_r is a weighting factor of the measured data $(Q_R)_r$, governed by the experimental conditions. This is a laborious and time-consuming method. Several authors have proposed successful ways of shortening the method[82,87,90].

The typical stages of such a numerical method were determined, for example, by Paoletti, Vacca, and Arenare,[81] who reinvestigated the two-step reaction Ag^+ + pyridine, this time considering the partial hydrolysis of the pyridine in aqueous solution. They used Sillén's "pit mapping method"[87-90] for their calculations. The results are included in Table 8.3. The system Ag^+/pyridine has also been studied by Izatt, Eatough, Snow, and Christensen,[82] using an iterative gradient method (VMM method). The results of their evaluation are also in Table 8.3.

Other examples of thermogram evaluation of multistep reactions are the following systems: $HgBr_2/Br^-$,[80] Cu^{2+}/pyridine,[82] Cd^{2+}/I^-,[91] $Hg(CN)_2$/thiourea,[92] Hg^{2+}/2-aminoethanol,[93] Cu^{2+}/1,10-phenanthroline,[93] and Zn^{2+}/1,10-phenanthroline.[93] The entropy titration with a strong acid or base as titrant is suitable only when the reaction has an equilibrium constant[94] of up to 10^3. Christensen et al.[96] have pointed out that equilibria for protonation reactions can be measured in a pK range 4 to 10 when weak acids or bases are used as titrants. If the concentrations of the reactants are chosen carefully, the accuracy of the thermodynamic data can be optimized.[95] Eatough[93] proposes a procedure of competitive equilibrium that makes it possible to obtain the constants of very stable ($K > 10^5$) metal-ligand complexes.

Thermometric titrations are primarily "enthalpy" titrations. This is the reason for the complicated experimental setup if both the enthalpy and the free enthalpy of a reaction are to be determined from the thermogram. If an enthalpy titration is combined with a "Gibbs energy" titration (e.g., potentiometry), which aims primarily at providing the equilibrium constant and thus ΔG^0, the evaluation is made much easier. In the iterations of (8.30), the K_i values are given quantities and only the function

$$U(\Delta H_{AB_i}) = \sum_{r=1}^{n} g_r [(Q_{R,\text{calc}})_r - (Q_{R,\text{exp}})_r]^2 \qquad (8.31)$$

must be minimized.

The combination of methods has frequently been used with success for determining thermodynamic data of multistep complex equilibria in aqueous solution.

Christensen et al. studied the halogen complexes in the multistep reaction $Hg^{2+} + X^- \rightarrow HgX_2(aq)$ (X = Cl, Br, I) and $HgX_2(s)$(X = I);[97] the stepwise formation of the complex ion $Hg(CN)_4^{2-}$ from Hg^{2+} and CN^-,[98] the systems $Zn^{2+} + CN^-$,[99] $Ag^+ + CN^-$,[100] $Co^{2+} + CN^-$,[101] and $Pd^{2+} + CN^-$,[102] the exchange reaction of the chloride ion in $HgCl_2(aq)$ with OH^-,[103] and with ethylenediamine, glycinate ion, and methylamine,[104] the stepwise dissociation of H_2L^+ (L = glycine, α-aminoisobutylric acid, threonine, or sarcosine); and the complex formation of L^- with Cu^{2+}.[105]

Gerding et al. report the complex formation in the systems $Cd^{2+} + X^-$ [X = F, Cl, Br, I];[106,107] $Cd^{2+} + SCN^-$ and $Cd^{2+} + CH_3COO^-$;[108] Cd^{2+} with CN^-, SCN^-, N_3^-, and NO_2^-;[109] and in the reaction of Cu^{2+}, Zn^{2+}, Cd^{2+}, and Pb^{2+} with CH_3COO^-.[110]

The investigations of Arnek et al. concern the hydrolysis equilibria with UO_2^{2+},[111] $Ni(ClO_4)_2$,[112] $Cd(ClO_4)_2$,[113] and $Hg(ClO_4)_2$.[114] The formation of the halogen complexes of Tl^{3+} has also been studied.[115,116]

Further data may be found in the comprehensive publication series of Christensen, Izatt, et al. on "Thermodynamics of Proton Dissociation in Aqueous Solution";[70-72,117-128] "Thermodynamics of Metal-Cyanide Coordination";[98-101,129-133] "Thermodynamics of Metal-Halide Coordination in Aqueous Solution";[97,134] or in that of Arnek et al., "Thermochemical Studies of Hydrolytic Reactions";[76,111-114,135-139] that of Gerding et al., "Thermochemical Studies of Metal Complexes";[106-110,140-143] and that of Grenthe et al., "Thermodynamic Properties of Rare Earth Complexes".[144-150] Other examples are the association of metal ions with cyclic polyethers;[151] mixed ligand complexes;[152,153] thermodynamics of Prussian blue and Turnbull's blue formation;[154,155] association of H^+, Cu^{2+},[156] and Ni^{2+},[157] with glycinate and phenylalanate ions; and lanthanide and actinide sulfate complexes.[158] Computer programs can be found in Refs. 86, 89, 90, 159, and 160.

Surveys of the field have been published by Christensen and Izatt[161] and by Hansen, Izatt, and Christensen;[162] the principles are revised in a recent series of articles by Christensen, Eatough, Izatt, and Ruckman;[163-165] a collection of data may be found in tables of Christensen and Izatt.[166, 167]

8.3 COMPLEXOMETRIC TITRATIONS

8.3.1 COMPLEX-ION FORMATION

Many of the complex-forming reactions discussed in the preceding sections are used as titration reactions. Examples of this type are the determination of zinc and copper in brass as $ZnHg(CNS)_4$ and $CuHg(CNS)_4$,[40] of zinc in ferrites[168] as $[Zn(CN)_4]^{2-}$ (cf. Ref. 2), and of copper in alloys as $[Cu(CN)_4]^{2-}$;[174] the formation of cyanide complexes is often a suitable titration reaction because it is fast and complete.[168-171] As the discussion in the preceding section has shown, the stoichiometrical course of complex-forming reactions is not always guaranteed; this must be taken into account when such reactions are chosen as titration reactions.

Comprehensive studies on the determination of fluoride, both by precipitation and complexation reactions, have been conducted by Everson and Ramirez[172] and by Deschamps, Deburck, and Bonnaire.[173] Everson and Ramirez found thorium, cerium, aluminum, and calcium to be suitable reactants for the thermometric titration of fluoride ion and discuss interference of various common ions in titration of F^- with different titrants. Deschamps et al. describe thermometric titrations with sodium fluoride of aluminum, iron, copper, lead, tin, and antimony salts; aluminum as $(AlF_6)^{3-}$, iron as $(FeF_6)^{3-}$; with less precision, copper as $(CuF_4)^{2-}$ can be determined. The destruction of fluoride complexes by boric acid also yields suitable titration curves, but aluminum and iron cannot be titrated in each other's presence. Sajó and Sipos used $(SnF_6)^{2-}$ for determination of tin in alloys;[174] $(SiF_6)^{2-}$ for silicon in steels, slags, clays, and minerals[168, 174-179] (also Ref. 180); and $(AlF_6)^{3-}$ for determining the aluminum content (see also Section 6.2.2) of these samples. Further examples of complex-ion formation are quoted in the preceding sections of this chapter, and in Chapters 4 and 6.

8.3.2 CHELATE FORMATION

The exchange of water molecules in the hydration shell of cations with other ligands takes place as a multistep reaction. More advantageous experimental conditions are obtained if the ligands offered to the metal cation are themselves combined in a molecule. Examples are the chelating agents, such as citrate and tartrate, titrated by Jordan and Ben-Yair[181] with zinc, cupric or cadmium nitrate; the oxalate, malonate, tartrate, and citrate complexes of aluminum(III) and iron(III) and the malonate, tartrate, and citrate complexes of gallium(III) have been investigated by Gallet and Pâris,[182,183] and the tartrate and citrate chelates of iron(III) also by Bobtelsky and Jordan.[184] Further chelating reactions are to be found in Section 8.1, for example, amino acid chelates, and in Section 8.3.3.

Beezer and Slawinski[185] use quinoxaline-2,3-dithiol in a DIE method for the determination of nickel and selenium.

8.3.3 COMPLEXES OF AMINOPOLYCARBOXYLIC ACIDS

In today's complexometry the aminopolycarboxylic acids are the preferred complex-forming agents; ethylenediaminetetraacetic acid (EDTA) is the most important example.[186]

Aminocarboxylates form complexes with the cations of the transition metals, the B metals (d^{10} cations) and A metals (d^0 cations), with the alkaline earth metals, and—to a lesser degree—also with Li^+ and Na^+. These complexes are soluble in water.

Different authors[187-189] calorimetrically studied the heats of reaction of EDTA complexes with mono- and divalent metal ions. The high stability of the EDTA chelate is explained by entropy effects. The following reaction was investigated:

$$M^{z+}(aq) + Y^{-4}(aq) \rightleftharpoons [MY]^{-4+z}(aq) + xH_2O$$

Part of the tetrasodium salt of EDTA exists in the form of NaY^{-3}, and part is hydrolyzed (HY^{-3}). At low salt concentrations, 1.4% of HY^{-3} is present and the necessary correction of the heat of reaction is 0.1 kcal mol^{-1}. The amount of correction decreases as the concentration is increased.

Thermodynamic data for dissociation and complex-forming reactions with divalent metal ions of other aminocarboxylates are also available.[186,190,191] Heats of formation for mixed-ligand chelates consisting of two different multidentate ligands linked to a central thorium(IV) ion have been studied thermometrically.[153] Pyrocatechol, tiron, chromotropic acid, potassium hydrogen phthalate, 8-hydroxyquinoline-5-sulfonic acid, iminodiacetic acid, 5-sulfosalicylic acid, and salicylic acid were used as the secondary ligands, while EDTA and DCTA (1,2-diaminocyclohexane-N,N,N',N'-tetraacetate) were used as primary ligands. The ΔH and ΔS values for the secondary ligand addition are determined.

TABLE 8.4. Thermodynamic Data and Accuracy of the End Point in Thermometric Titrations with EDTA

Cation	$\log K$	Reference 192		Reference 196[a]	Reference 193	References 198 and 199
		ΔH^0 kcal mol^{-1}	EP %	ΔH^0 kcal mol^{-1}	EP %	EP %
Bi^{3+}	27.9	—	—	—	—	<1
Fe^{3+}	25.1	—	—	—	c	—
In^{3+}	24.9	—	—	—	—	<1
Hg^{2+}	21.8	—	—	—	c	<1
Sn^{2+}	22.1	—	—	—	—	<1
Cu^{2+}	18.8	−8.2	—	−8.3[a]	0.5	<1
Ni^{2+}	18.6	−7.4	0.5	−7.2[a]	0.6	<1
Pd^{2+}	18.5	—	—	—	—	<1
Pb^{2+}	18.0	−12.8	1.0	−12.7[a]	—	<1
Cd^{2+}	16.5	−9.2	0.4	−9.7	0.5	<1
Zn^{2+}	16.5	−4.6	0.8	−4.5	0.4	<1
Co^{2+}	16.1	−4.2	0.1	−3.5	—	<1
Al^{3+}	16.1	—	—	+10.9	0.8	—
Ce^{3+}	16.0	—	—	—	0.9	—
La^{3+}	15.5	—	—	−3.9	—	—
Fe^{2+}	14.3	—	—	—	c	—
Mn^{2+}	13.8	—	—	−6.4[a]	c	—
Ca^{2+}	10.7	−5.7	1.0	−5.6	1.0	<1
Be^{2+}	9.0	—	—	+2.3	1.4	—
Mg^{2+}	8.7	+5.5	0.4	+4.8	1.6	<1
Sr^{2+}	8.6	—	—	−4.9	c	<1
Ba^{2+}	7.8	—	—	−4.6[a]	0.9	<1
Li^+	2.8	—	—	+2.2	b	—
Na^+	—	—	—	—	—	—
K^+	—	—	—	+2.2	—	—
NH_4^+	—	—	—	+4.8	b	—
Ag^+	—	—	—	—	2.0	—
Ga^{3+}	—	—	—	—	—	<1
Cr^{3+}	—	—	—	+7.3	0.2	—
Co^{3+}	—	—	—	—	0.3	—
Sn^{4+}	—	—	—	—	0.6	—
Th^{4+}	—	—	—	—	—	<1

[a] Error analysis for end-point accuracy (EP < 1%) in Ref. 196 has been attempted for representative examples only Cu^{2+}, Ni^{2+}, Pb^{2+}, Mn^{2+} and Ba^{2+}.
[b] Unsuccessful titration; no end point indicated because of insufficient heat of reaction.[193]
[c] Indicated in Ref. 193: No results

Complexometric thermometric titrations with EDTA have been conducted with every thermometric method, the TET method,[192,193] the DIE method,[194] thermometric-microanalytically,[195] and with the continuous flow method.[196] Freeberg[197] used NTA (nitrilotriacetic acid) as the complexing agent in chelation reactions with Cu(II), Zn(II), Ni(II), and Co(II).

Weisz and Kiss[198] and Kiss[199] used catalytic thermometric indication[200] of the end point (see Section 5.5.1) by back titration with standard manganese(II) solution in titrations of many metal cations with EDTA, DCTA, and NTA in water and dimethyl sulfoxide. The decomposition of hydrogen peroxide and the reaction between hydrogen peroxide and resorcinol,[200] both catalyzed by manganese, have been utilized as indicator systems. In these titrations, the end point is yielded with the accuracy characteristic of the catalytic thermometric method, even at the microgram level. Masking can be employed in the titration in DMSO in order to determine selectively individual types of ions.

Table 8.4 gives a survey of titrations reported in Refs. 192, 193, 196, 198, and 199. The log K values are taken from Refs. 186 and 195. In Refs. 194 and 195, only titrations of Pb^{2+} and Mg^{2+} are given as examples of the applicability of the method described in these investigations. These two values fit well into the table and are, therefore, not listed separately.

Table 8.4 shows that some reactions are endothermic, others exothermic. How to profit from this fact has been pointed out above (Fig. 1.2). The titration of Ca^{2+} (exothermic) alongside Mg^{2+} (endothermic) is discussed in Refs. 192 and 201; other examples of this type are discussed by Bark and Bark.[202]

De Leo and Stern[203] compare the classical and thermometric methods for titration of $MgCl_2$ with EDTA and prefer the latter. Compared with neutralization or redox reactions, many complex reactions have small heats of reaction; therefore, temperature must be measured with high accuracy. Alleman[204] states that an accuracy of 3% is possible at a concentration of 5×10^{-4} M, but this statement is made obsolete by modern techniques[205] and by the study of Weisz and Kiss.[198,199]

REFERENCES

1. P. Mondain-Monval and R. Pâris, *Bull. Soc. Chim. Fr.*, **5**, 1641 (1938).
2. P. Mondain-Monval and R. Pâris, *C. R. Acad. Sci. (Paris)*, **198**, 1154 (1934).
3. R. Pâris, *C. R. Acad. Sci. (Paris)*, **199**, 863 (1934).
4. J. Barthel, N. G. Schmahl, and K. Lenz, *Z. Anal. Chem.*, **233**, 328 (1968).
5. B. C. Haldar, *J. Indian Chem. Soc.*, **23**, 147 (1946).
6. B. C. Haldar, *J. Indian Chem. Soc.*, **23**, 153 (1946).
7. B. C. Haldar, *J. Indian Chem. Soc.*, **23**, 183 (1946).
8. B. C. Haldar, *J. Indian Chem. Soc.*, **25**, 439 (1948).

9. B. C. Haldar, *J. Indian Chem. Soc.*, **25**, 445 (1948).
10. B. C. Haldar, *J. Indian Chem. Soc.*, **23**, 205 (1946).
11. B. C. Haldar, *J. Indian Chem. Soc.*, **27**, 484 (1950).
12. B. C. Haldar, *J. Indian Chem. Soc.*, **24**, 503 (1947).
13. B. C. Purkayastha, *J. Indian Chem. Soc.*, **24**, 257 (1947).
14. B. C. Purkayastha and R. N. Sen-Sarma, *J. Indian Chem. Soc.*, **23**, 31 (1946).
15. S. Banerjee, *J. Indian Chem. Soc.*, **27**, 417 (1950).
16. A. K. Bhattacharya and H. C. Gaur, *J. Indian Chem. Soc.*, **24**, 487 (1947).
17. A. K. Bhattacharya and H. C. Gaur, *J. Indian Chem. Soc.*, **25**, 185 (1948).
18. H. C. Gaur and A. K. Bhattacharya, *J. Indian Chem. Soc.*, **26**, 46 (1949).
19. H. C. Gaur and A. K. Bhattacharya, *Proc. Nat. Acad. Sci., India*, Sect. A., **19**, 45 (1950).
20. H. C. Gaur and A. K. Bhattacharya, *J. Indian Chem. Soc.*, **27**, 131 (1950).
21. R. S. Saxena and A. K. Bhattacharya, *J. Indian Chem. Soc.*, **28**, 703 (1951).
22. A. K. Bhattacharya and R. S. Saxena, *J. Indian Chem. Soc.*, **29**, 529 (1952).
23. A. K. Bhattacharya and R. S. Saxena, *J. Indian Chem. Soc.*, **29**, 263 (1952).
24. H. C. Gaur and A. K. Bhattacharya, *J. Indian Chem. Soc.*, **29**, 29 (1952).
25. H. C. Gaur and A. K. Bhattacharya, *J. Indian Chem. Soc.*, **29**, 117 (1952).
26. R. S. Saxena and A. K. Bhattacharya, *J. Indian Chem. Soc.*, **29**, 632 (1952).
27. H. C. Gaur and A. K. Bhattacharya, *Sci. Cult.*, **20**, 237 (1954).
28. R. S. Saxena, *J. Indian Chem. Soc.*, **31**, 56 (1954).
29. J. N. Gaur, H. C. Gaur, and A. K. Bhattacharya, *J. Indian Chem. Soc.*, **30**, 859 (1953).
30. J. N. Gaur and M. S. Bhadraver, *J. Indian Chem. Soc.*, **36**, 108 (1959).
31. N. K. Dutt, *J. Indian Chem. Soc.*, **22**, 97 (1945).
32. N. K. Dutt, *J. Indian Chem. Soc.*, **22**, 107 (1945).
33. D. G. Vartak and M. B. Kabadi, *J. Indian Chem. Soc.*, **32**, 351 (1955).
34. D. G. Vartak and M. B. Kabadi, *J. Univ. Bombay*, **22**, Part 5, 34 (1954).
35. M. R. Nayar and C. S. Pande, *J. Indian Chem. Soc.*, **28**, 112 (1951).
36. M. Bose and D. M. Chowdhury, *J. Indian Chem. Soc.*, **32**, 673 (1955).
37. B. C. Haldar, *Nature*, **166**, 744 (1950).
38. S. Banerjee and B. C. Haldar, *Nature*, **165**, 1012 (1950).
39. K. K. Chatterji, *J. Indian Chem. Soc.*, **32**, 366 (1955).
40. K. K. Chatterji, *J. Indian Chem. Soc.*, **35**, 57 (1958).
41. K. K. Chatterji and A. K. Ghosh, *J. Indian Chem. Soc.*, **34**, 407 (1957).
42. K. K. Chatterji, *J. Indian Chem. Soc.*, **35**, 709 (1958).
43. K. K. Chatterji, *J. Indian Chem. Soc.*, **35**, 883 (1958).
44. P. Job, *C. R. Acad. Sci. (Paris)*, **180**, 928 (1925).
45. P. Job, *Ann. Chim.*, **9**, 113 (1928).
46. P. Hagenmuller, *C. R. Acad. Sci. (Paris)*, **230**, 2190 (1950).
47. E. Asmus, *Z. Anal. Chem.*, **183**, 321 (1961).
48. E. Asmus, *Z. Anal. Chem.*, **183**, 401 (1961).
49. E. Asmus and P. Meyer, *Z. Anal. Chem.*, **190**, 390 (1962).
50. Chauvenet, P. Job, and G. Urbain, *C. R. Acad. Sci. (Paris)*, **171**, 855 (1920).
51. S. K. Siddhanta, *J. Indian Chem. Soc.*, **25**, 579 (1948).
52. S. K. Siddhanta and M. P. Guha, *J. Indian Chem. Soc.*, **32**, 355 (1955).
53. M. S. Bhadraver and J. N. Gaur, *J. Indian Chem. Soc.*, **36**, 103 (1959).

54. S. T. Zenchelsky and P. R. Segatto, *J. Am. Chem. Soc.*, **80**, 4796 (1958).
55. F. J. Cioffi and S. T. Zenchelsky, *J. Phys. Chem.*, **67**, 357 (1963).
56. M. H. Dilke and D. D. Eley, *J. Chem. Soc.*, **1949**, 2601.
57. P. Papoff and P. G. Zambonin, *Ric. Sci.*, **35**, 93 (1965).
58. J. J. Christensen, R. M. Izatt, L. D. Hansen, and J. A. Partridge, *J. Phys. Chem.*, **70**, 2003 (1966).
59. L. D. Hansen, J. J. Christensen, and R. M. Izatt, *Chem. Commun.*, **1965**, 36.
60. R. M. Izatt, W. C. Fernelius, and B. P. Block, *J. Phys. Chem.*, **59**, 235 (1955).
61. I. Poulsen and J. Bjerrum, *Acta Chem. Scand.*, **9**, 1407 (1955).
62. S. Hietanen and L. G. Sillén, *Acta Chem. Scand.*, **8**, 1607 (1954).
63. K. Schlyter and L. G. Sillén, *Acta Chem. Scand.*, **13**, 385 (1959).
64. K. Schlyter, *Trans. R. Inst. Technol. (Stockh.)*, **132** (1959).
65. K. Schlyter, *Trans. R. Inst. Technol. (Stockh.)*, **152** (1960).
66. K. Schlyter and D. L. Martin, *Trans. R. Inst. Technol. (Stockh.)*, **175** (1961).
67. K. Schlyter, *Trans. R. Inst. Technol. (Stockh.)*, **182** (1961).
68. F. Becker and R. Grundmann, *Z. Phys. Chem.* (New Series), **66**, 137 (1969).
69. J. Barthel, F. Becker, and N. G. Schmahl, *Z. Phys. Chem.* (New Series), **29**, 58 (1961).
70. R. M. Izatt, J. H. Rytting, L. D. Hansen, and J. J. Christensen, *J. Am. Chem. Soc.*, **88**, 2641 (1966).
71. J. J. Christensen, J. H. Rytting, and R. M. Izatt, *J. Phys. Chem.*, **71**, 2700 (1967).
72. J. J. Christensen, J. H. Rytting, and R. M. Izatt, *J. Am. Chem. Soc.*, **88**, 5105 (1966).
73. D. P. Wrathall, R. M. Izatt, and J. J. Christensen, *J. Am. Chem. Soc.*, **86**, 4779 (1964).
74. R. J. Raffa, M. J. Stern, and L. Malspeis, *Anal. Chem.*, **40**, 70 (1968).
75. L. D. Hansen, B. D. West, E. J. Baca, and C. L. Blank, *J. Am. Chem. Soc.*, **90**, 6588 (1968).
76. R. Arnek and C. C. Patel, *Acta Chem. Scand.*, **22**, 1097 (1968).
77. F. M. Arshid, C. H. Giles, E. C. McLure, A. Ogilivie, T. J. Rose, and J. C. Eaton, *J. Chem. Soc.*, **1955**, 67.
78. M. D. Joesten and R. S. Drago, *J. Am. Chem. Soc.*, **84**, 2696 (1962).
79. F. Becker, J. Barthel, N. G. Schmahl, G. Lange, and H. M. Lüschow, *Z. Phys. Chem.* (New Series), **37**, 33 (1963).
80. F. Becker, J. Barthel, N. G. Schmahl, and H. M. Lüschow, *Z. Phys. Chem.* (New Series), **37**, 52 (1963).
81. P. Paoletti, A. Vacca, and D. Arenare, *Coordin. Chem. Rev.*, **1**, 280 (1966).
82. R. M. Izatt, D. Eatough, R. L. Snow, and J. J. Christensen, *J. Phys. Chem.*, **72**, 1208 (1968).
83. R. K. Murmann and F. Basolo, *J. Am. Chem. Soc.*, **77**, 3484 (1955).
84. W. C. Vosburgh and S. A. Cogswell, *J. Am. Chem. Soc.*, **65**, 2412 (1943).
85. F. Becker, *Chem.-Ing. Tech.*, **41**, 1105 (1969).
86. R. Arnek, *Ark. Kemi*, **32**, 81 (1970).
87. L. G. Sillén, *Acta Chem. Scand.*, **16**, 159 (1962).
88. L. G. Sillén, *Acta Chem. Scand.*, **18**, 1085 (1964).
89. N. Ingri and L. G. Sillén, *Acta Chem. Scand.*, **16**, 173 (1962).

90. N. Ingri and L. G. Sillén, *Ark. Kemi*, **23**, 97 (1964).
91. F. Becker and H. M. Lüschow, *Proc. 8th Int. Conf. Coord. Chem.*, Vienna, 1964, p. 334.
92. R. M. Izatt, D. Eatough, and J. J. Christensen, *J. Phys. Chem.*, **72**, 2720 (1968).
93. D. J. Eatough, *Anal. Chem.*, **42**, 635 (1970).
94. L. D. Hansen, J. J. Christensen, and R. M. Izatt, *Chem. Commun.*, **3**, 36 (1965).
95. J. J. Christensen, D. P. Wrathall, J. O. Oscarson, and R. M. Izatt, *Anal. Chem.*, **40**, 1713 (1968).
96. J. J. Christensen, D. P. Wrathall, and R. M. Izatt, *Anal. Chem.*, **40**, 175 (1968).
97. J. J. Christensen, R. M. Izatt, L. D. Hansen, and J. D. Hale, *Inorg. Chem.*, **3**, 130 (1964).
98. J. J. Christensen, R. M. Izatt, and D. Eatough, *Inorg. Chem.*, **4**, 1278 (1965).
99. R. M. Izatt, J. J. Christensen, J. W. Hansen, and G. D. Watt, *Inorg. Chem.*, **4**, 718 (1965).
100. R. M. Izatt, H. D. Johnston, G. D. Watt, and J. J. Christensen, *Inorg. Chem.*, **6**, 132 (1967).
101. R. M. Izatt, G. D. Watt, C. H. Bartholomew, and J. J. Christensen, *Inorg. Chem.*, **7**, 2236 (1968).
102. G. D. Watt, D. Eatough, R. M. Izatt, and J. J. Christensen, *Proc. Utah Acad. Sci., Arts Lett.*, **42**, 298 (1965).
103. J. A. Partridge, R. M. Izatt, and J. J. Christensen, *J. Chem. Soc.*, **1965**, 4231.
104. J. A. Partridge, J. J. Christensen, and R. M. Izatt, *J. Am. Chem. Soc.*, **88**, 1649 (1966).
105. R. M. Izatt, J. J. Christensen, and V. Kothari, *Inorg. Chem.*, **3**, 1565 (1964).
106. P. Gerding, *Acta Chem. Scand.*, **20**, 79 (1966).
107. P. Gerding and I. Jönsson, *Acta Chem. Scand.*, **22**, 2247 (1968).
108. P. Gerding and B. Johansson, *Acta Chem. Scand.*, **22**, 2255 (1968).
109. P. Gerding, *Acta Chem. Scand.*, **20**, 2771 (1966).
110. P. Gerding, *Acta Chem. Scand.*, **21**, 2015 (1967).
111. R. Arnek and K. Schlyter, *Acta Chem. Scand.*, **22**, 1331 (1968).
112. R. Arnek and W. Kakolowicz, *Acta Chem. Scand.*, **21**, 2180 (1967).
113. R. Arnek and W. Kakolowicz, *Acta Chem. Scand.*, **21**, 1449 (1967).
114. R. Arnek and I. Szilárd, *Acta Chem. Scand.*, **22**, 1334 (1968).
115. I. Leden and T. Ryhl, *Acta Chem. Scand.*, **18**, 1196 (1964).
116. I. Grenthe and I. Leden, *Proc. 8th Int. Conf. Coord. Chem.*, Vienna, 1964, p. 332.
117. R. M. Izatt and J. J. Christensen, *J. Phys. Chem.*, **66**, 359 (1962).
118. J. J. Christensen and R. M. Izatt, *J. Phys. Chem.*, **66**, 1030 (1962).
119. L. D. Hansen, J. A. Partridge, R. M. Izatt, and J. J. Christensen, *Inorg. Chem.*, **5**, 569 (1966).
120. J. J. Christensen, R. M. Izatt, and L. D. Hansen, *J. Am. Chem. Soc.*, **89**, 213 (1967).
121. J. J. Christensen, D. P. Wrathall, R. M. Izatt, and D. O. Tolman, *J. Phys. Chem.*, **71**, 3001 (1967).

122. J. J. Christensen, J. L. Oscarson, and R. M. Izatt, *J. Am. Chem. Soc.*, **90**, 5949 (1968).
123. J. J. Christensen, R. M. Izatt, D. P. Wrathall, and L. D. Hansen, *J. Chem. Soc., A*, **1969**, 1212.
124. J. J. Christensen, H. D. Johnston, and R. M. Izatt, *J. Chem. Soc., A*, **1970**, 454.
125. J. J. Christensen, M. D. Slade, D. E. Smith, R. M. Izatt, and J. Tsang, *J. Am. Chem. Soc.*, **92**, 4164 (1970).
126. J. J. Christensen, J. H. Rytting, and R. M. Izatt, *J. Chem. Soc., B*, **1970**, 1643.
127. J. J. Christensen, J. H. Rytting, and R. M. Izatt, *J. Chem. Soc., B*, **1970**, 1646.
128. J. J. Christensen, D. E. Smith, M. D. Slade, and R. M. Izatt, *Thermochim. Acta*, **5**, 35 (1972/73).
129. R. M. Izatt, J. J. Christensen, R. T. Pack, and R. Bench, *Inorg. Chem.*, **1**, 828 (1962).
130. J. J. Christensen, R. M. Izatt, J. D. Hale, R. T. Pack, and G. D. Watt, *Inorg. Chem.*, **2**, 337 (1963).
131. G. D. Watt, J. J. Christensen, and R. M. Izatt, *Inorg. Chem.*, **4**, 220 (1965).
132. R. M. Izatt, G. D. Watt, D. Eatough, and J. J. Christensen, *J. Chem. Soc., A*, **1967**, 1304.
133. R. M. Izatt, H. D. Johnston, D. J. Eatough, J. W. Hansen, and J. J. Christensen, *Thermochim. Acta*, **2**, 77 (1971).
134. L. D. Hansen, R. M. Izatt, and J. J. Christensen, *Inorg. Chem.*, **2**, 1243 (1963).
135. R. Arnek, *Acta Chem. Scand.*, **22**, 1102 (1968).
136. R. Arnek and K. Schlyter, *Acta Chem. Scand.*, **22**, 1327 (1968).
137. R. Arnek, *Acta Chem. Scand.*, **23**, 1986 (1969).
138. R. Arnek and L. Barcza, *Acta Chem. Scand.*, **26**, 213 (1972).
139. R. Arnek and S. R. Johansson, *Acta Chem. Scand.*, **26**, 2903 (1972).
140. P. Gerding, *Acta Chem. Scand.*, **20**, 2624 (1966).
141. P. Gerding, *Acta Chem. Scand.*, **21**, 2007 (1967).
142. P. Gerding, *Acta Chem. Scand.*, **22**, 1283 (1968).
143. P. Gerding, *Acta Chem. Scand.*, **23**, 1695 (1969).
144. I. Grenthe, *Acta Chem. Scand.*, **17**, 2487 (1963).
145. I. Grenthe, *Acta Chem. Scand.*, **18**, 283 (1964).
146. I. Grenthe and D. R. Williams, *Acta Chem. Scand.*, **21**, 347 (1967).
147. I. Grenthe and E. Hansson, *Acta Chem. Scand.*, **23**, 611 (1969).
148. I. Grenthe and G. Gårdhammar, *Acta Chem. Scand.*, **26**, 3207 (1972).
149. I. Grenthe and H. Ots, *Acta Chem. Scand.*, **26**, 1229 (1972).
150. I. Dellien, *Acta Chem. Scand.*, **27**, 733 (1973).
151. R. M. Izatt, D. P. Nelson, J. H. Rytting, B. L. Haymore, and J. J. Christensen, *J. Am. Chem. Soc.*, **93**, 1619 (1971).
152. A. Yingst, R. M. Izatt, and J. J. Christensen, *J. Chem. Soc., A*, **1972**, 1199.
153. G. C. Kugler and G. H. Carey, *Talanta*, **17**, 907 (1970).
154. G. D. Watt, *Diss. Abstr.*, **27B**, (1966), 1406 (order No. 66-10, 520).
155. R. M. Izatt, G. D. Watt, C. H. Bartholomew, and J. J. Christensen, *Inorg. Chem.*, **9**, 2019 (1970).

156. K. P. Anderson, W. O. Greenhalgh, and R. M. Izatt, *Inorg. Chem.*, **5**, 2106 (1966).
157. K. P. Anderson, W. O. Greenhalgh, and F. A. Butler, *Inorg. Chem.*, **6**, 1056 (1967).
158. R. G. de Carvalho and G. R. Choppin, *J. Inorg. Nucl. Chem.*, **29**, 737 (1967).
159. R. Arnek, L. G. Sillén, and O. Wahlberg, *Ark. Kemi*, **31**, 353 (1969).
160. P. Brauner, L. G. Sillén, and R. Whiteker, *Ark. Kemi*, **31**, 365 (1969).
161. J. J. Christensen and R. M. Izatt, in H. A. O. Hill and P. Day, *Physical Methods in Advanced Inorganic Chemistry*, Interscience, New York, 1968 (reprinted 1970), p. 538.
162. L. D. Hansen, R. M. Izatt, and J. J. Christensen, in J. Jordan, *New Developments in Titrimetry*, Vol. 2, Dekker, New York, 1974.
163. J. J. Christensen, J. Ruckman, D. J. Eatough, and R. M. Izatt, *Thermochim. Acta*, **3**, 203 (1971/72).
164. D. J. Eatough, J. J. Christensen, and R. M. Izatt, *Thermochim. Acta*, **3**, 219 (1971/72).
165. D. J. Eatough, R. M. Izatt, and J. J. Christensen, *Thermochim. Acta*, **3**, 233 (1971/72).
166. J. J. Christensen and R. M. Izatt, *Handbook of Metal Ligand Heats and Related Thermodynamic Quantities*, Dekker, New York, 1970.
167. R. M. Izatt and J. J. Christensen, "Heats of Proton Ionization, pK and Related Thermodynamic Quantities," in H. A. Sober, *Handbook of Biochemistry*, 2nd ed., Chemical Rubber Co., Cleveland, Ohio, 1970, p. J-58.
168. I. Sajó and B. Sipos, *Z. Anal. Chem.*, **222**, 23 (1966).
169. J. L. Rasmussen and T. Nielsen, *Acta Chem. Scand.*, **17**, 1623 (1963).
170. G. Rády, E. Kaiser, and O. Gimesi, *Period. Polytech.*, **11**, 111 (1967).
171. I. Sajó and B. Sipos, *Talanta*, **14**, 203 (1967).
172. W. L. Everson and E. M. Ramirez, *Anal. Chem.*, **39**, 1771 (1967).
173. P. Deschamps, A. Deburck, and Y. Bonnaire, *Anal. Chim. Acta*, **40**, 259 (1968).
174. I. Sajó and A. Sipos, *Bányász. és Kohász. Lapok*, **101**, 484 (1968).
175. I. Sajó, *Kohász. Lapok*, **7**, 287 (1957).
176. I. Sajó and J. Ujvári, *Z. Anal. Chem.*, **202**, 177 (1964).
177. I. Sajó, *Z. Anal. Chem.*, **242**, 165 (1968).
178. I. Sajó and B. Sipos, *Zem.-Kalk-Gips*, **21**, 32 (1968).
179. I. Sajó and B. Sipos, *Radex-Rundsch.*, **1968**, 178.
180. M. Mandl, J. Scála, and M. Kaše, *Neue Hütte*, **7**, 176 (1962).
181. J. Jordan and M. P. Ben-Yair, *Ark. Kemi*, **11**, 239 (1956).
182. J. P. Gallet and R. A. Pâris, *Anal. Chim. Acta*, **39**, 181 (1967).
183. J. P. Gallet and R. A. Pâris, *Anal. Chim. Acta*, **39**, 341 (1967).
184. M. Bobtelsky and J. Jordan, *J. Am. Chem. Soc.*, **69**, 2286 (1947).
185. A. E. Beezer and A. K. Slawinski, *Talanta*, **18**, 837 (1971).
186. G. Schwarzenbach and H. Flaschka, *Die komplexometrische Titration*, 5th ed., Ferdinand Enke Verlag, Stuttgart, 1965.
187. R. G. Charles, *J. Am. Chem. Soc.*, **76**, 5854 (1954).
188. R. A. Care and L. A. K. Staveley, *J. Chem. Soc.*, **1956**, 4571.

189. A. P. Brunetti, G. H. Nancollas, and P. N. Smith, *J. Am. Chem. Soc.*, **91**, 4680 (1969).
190. G. Degischer and G. H. Nancollas, *Inorg. Chem.*, **9**, 1259 (1970).
191. S. Boyd, A. Bryson, G. H. Nancollas, and K. Torrance, *J. Chem. Soc.*, **1965**, 7353.
192. J. Jordan and T. G. Alleman, *Anal. Chem.*, **29**, 9 (1957).
193. P. T. Priestley, *Analyst*, **88**, 194 (1963).
194. J. C. Wasilewski, P. T.-S. Pei, and J. Jordan, *Anal. Chem.*, **36**, 2131 (1964).
195. J. Jordan, R. A. Henry, and J. C. Wasilewski, *Microchem. J.*, **10**, 260 (1966).
196. P. T. Priestley, W. S. Sebborn, and R. F. W. Selman, *Analyst*, **90**, 589 (1965).
197. F. E. Freeberg, *Anal. Chem.*, **41**, 54 (1969).
198. H. Weisz and T. Kiss, *Z. Anal. Chem.*, **249**, 302 (1970).
199. T. Kiss, *Z. Anal. Chem.*, **252**, 12 (1970).
200. H. Weisz and T. Janjić, *Z. Anal. Chem.*, **227**, 1 (1967).
201. J. Jordan, *Chimia*, **17**, 101 (1963).
202. L. S. Bark and S. M. Bark, *Thermometric Titrimetry*, Pergamon, Oxford, 1969.
203. A. B. De Leo and M. J. Stern, *J. Pharm. Sci.*, **54**, 911 (1965).
204. T. G. Alleman, Abstr. 132nd Meeting, Am. Chem. Soc., New York, **1957**, 11B.
205. R. Wachter, J. Barthel, and K. Wachter-Lenz, in preparation.

CHAPTER

9

INSTRUMENTATION IN TITRATION CALORIMETRY

R. WACHTER,
University of Regensburg

The basic design of titration calorimeters was described in Chapters 2 and 3. The following sections discuss the principles to be observed in constructing suitable titration calorimeters for different fields of application. The different parts of the basic equation of thermometric titration given in Chapter 2, which includes the different values necessary for an evaluation, will be treated separately, as well as the problems of methodology defined in Chapter 3.

9.1 PRINCIPLES OF DESIGN

9.1.1 HEAT EXCHANGE BETWEEN THE CALORIMETER AND ITS ENVIRONMENT

The system consisting of calorimeter vessel and its content is continually exchanging heat and material with its environment. The former is always considered in the basic equation, the latter only in cases where it is intended by the method used (e.g., flow calorimetry, overflow calorimeter). An uncontrolled exchange of materials with special heat effect can also arise, for example, from losses caused by evaporation and must be counteracted by a suitable construction of the calorimeter.

The heat exchange between the calorimeter and its environment is respected in a general way in the basic equation of thermometric titration by (2.16) as a form of Newton's law of cooling

$$w_A = -\kappa C(T-\Theta) \tag{9.1a}$$

To keep the heat exchange low, a small constant of exchange κ and a small difference of temperature $(T-\Theta)$ between calorimeter vessel and environment must be aimed at.

In order to find out the best calorimeter construction, the overall quantity κ must be split into its different components, namely, those associated with heat transfer by radiation, by conduction, and by convection.

These several parts are discussed in the following sections and their contribution to κ demonstrated using the calorimeter sketched in Figure 9.1.

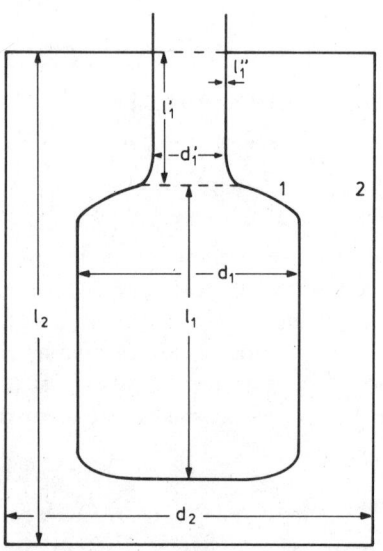

Fig. 9.1. Sketch of a calorimeter vessel (1) and its enclosure (2). The calculations in the text are based on the following dimensions: $d_1 = 6$ cm, $d_1' = 2.2$ cm, $l_1 = 8$ cm, $l_1' = 3$ cm, $l_1'' = 0.2$ cm, $d_2 = 14$ cm, $l_2 = 14$ cm.

Heat Transfer by Radiation. The calorimeter vessel (1), Fig. 9.1, at temperature T(K) exchanges heat by radiation with its enclosure (2), which has a temperature Θ(K). The power of the radiation entering the vessel $w_A^{(r)}$ (W) is given by

$$w_A^{(r)} = -\sigma A_1 F_{12}(T^4 - \Theta^4) \qquad (9.2)$$

In (9.2) σ is the Stefan–Boltzmann constant ($\sigma = 5.67 \cdot 10^{-8}$ W m^{-2} K^{-4}), A_1 (m^2) is the surface area of the calorimeter vessel, and F_{12} (dimensionless) is the emissivity factor, which is defined as a function of radiation emissivities, ϵ_1, ϵ_2, and the geometry of the surfaces A_1, A_2 exchanging heat. For our case of small differences of temperature, it is sufficient to consider the first approximation of (9.2)

$$w_A^{(r)} = -4\sigma A_1 F_{12} T^3 (T - \Theta) \qquad (9.3)$$

At a mean temperature $T = 300$ K and a temperature difference $\Delta T = 1$K, the error in (9.3) compared with (9.2) is less than 0.01%.

The emissivity factor F_{12} for the calorimeter depicted in Fig. 9.1 is that of a completely enclosed body differing in size and shape from its enclosure.[1]

$$F_{12} = \left[\frac{1}{\epsilon_1} + \frac{A_1}{A_2}\left(\frac{1}{\epsilon_2} - 1\right)\right]^{-1} \tag{9.4}$$

TABLE 9.1. Emissivities of Some Materials

Surface	Emissivity ϵ
Gold (polished)	0.018 ($t = 130°C$)
Silver (smooth)	0.022 ($t = 20°C$)
Copper (polished, unoxidized)	0.03 ($t = 20°C$)
Copper (oxidized, black)	0.78 ($t = 20°C$)
Nickel (polished)	0.045 ($t = 100°C$)
Chromium (polished)	0.06 ($t = 150°C$)
Glass	0.94 ($t = 90°C$)

Table 9.1 contains the emissivities ϵ of a number of materials used in constructing calorimeters.[2] As can be seen from the table, polished metals generally have low emissivities. In practical work, therefore, the high emissivity of glass or the base metals is reduced by plating with nonoxidizing metals.

By using (9.3) in combination with (9.4) and the data from Table 9.1, we obtain $F_{12} = 0.018$ and $w_A^{(r)} = 6 \cdot 10^{-5}$ cal sec^{-1} for the calorimeter in Fig. 9.1. It is assumed that the calorimeter vessel is of glass, cylindrically shaped and silver-plated, enclosed in a nickel-plated brass cylinder; and that $\Delta T = 0.1$ K at $T = 300$ K. The calculated radiation power of $6 \cdot 10^{-5}$ cal sec^{-1} with a calorimeter heat capacity of about 200 cal K^{-1} results in a rate of temperature change of $2 \cdot 10^{-5}$ K min^{-1}.

Heat Transfer by Conduction. The thermal conductivity λ(W m^{-1} K^{-1}) brings about a rate of heat flow according to

$$w_A^{(c)}(x) = -\lambda A \left(\frac{\partial T}{\partial x}\right), \tag{9.5}$$

in which the sectional area A (m^2) is perpendicular to the x direction. Table 9.2 contains the values for λ of solid and gaseous materials.[2]

TABLE 9.2. Thermal Conductivities at 20°C

Material	λ W m^{-1} K^{-1}	Material	λ W m^{-1} K^{-1}
Silver	428	Stainless steel (18 Cr/Ni)	14.5
Copper	395	Silica	1.36
Gold	312	Pyrex	1.13
Aluminum	239	Styrofoam (20 kg m^{-3})	0.035
Brass (70% Cu)	112	Air	0.025

Table 9.2 shows that it is impossible to achieve complete isolation, as can be done electrically. The ratio of thermal conductivities of the best conductor (silver) and the best insulator (Styrofoam) is 10^4; the corresponding ratio of electrical conductivities is 10^{25}. It is advisable therefore, to avoid heat transfer by conduction and place the calorimeter vessel in an evacuated enclosure instead.

If the mean free path of the gas molecules is long relative to the distance between the calorimeter vessel and its enclosure, the rate of heat flow $w_A^{(c)}$ (gas) between the calorimeter and its environment is

$$w_A^{(c)}(\text{g}) = -\eta p A \sqrt{\frac{273}{T}} \, (T-\Theta) \qquad (9.6)$$

η is the viscosity of the gas (P), p is the pressure (mm Hg), and A (cm^2) is the surface area of the calorimeter vessel.

In the calorimeter of Fig. 9.1, the support of the calorimeter vessel (Pyrex tube) is cemented into the brass lid in a heat-conducting way. According to (9.5), using the data for Pyrex from Table 9.2, with $\Delta T = 0.1$ K and an effective sectional area for heat conduction of $1.4 \cdot 10^{-4}$ m^2, we obtain a rate of heat flow $w_A^{(c)} = 1.3 \cdot 10^{-4}$ cal sec^{-1} and a rate of temperature change of $4 \cdot 10^{-5}$ K min^{-1} (heat capacity of the calorimeter: 200 cal K^{-1}). To this must be added the heat transfer due to gaseous conduction. At a pressure of 10^{-4} mm Hg and with $\eta = 1.8 \cdot 10^{-4}$ P (air), $\Delta T = 0.1$ K and $T = 300$ K, (9.6) gives a value for $w_A^{(c)}$ (gas) of 10^{-7} cal sec^{-1}. This can be neglected in comparison with the part played by the heat transfer of the solid bodies.

Heat Transfer by Convection. The heat transfer caused by convection of the air between the calorimeter vessel and its enclosure may be neglected in all those cases in which the space around the calorimeter vessel is evacuated or filled with solid insulating material. Heat transfer by convection decreases with the square of the gas pressure and becomes negligible in the range of few millimeters of mercury.

The Overall Cooling Constant κ. The preceding discussion has shown that all the different rates of heat transfer that must be considered are proportional to $T - \Theta$ at a first approximation, provided that the temperature difference $\Delta T = T - \Theta$ is small. This is the basis of the general law of (9.1a). Only the magnitude of κ is relevant for the application of the calorimeter equation. A simple way of determining κ consists in measuring the differential coefficient dT/dt or the quotient of the differences $\Delta T/\Delta t$ when the temperature difference $(T - \Theta)$ is known.[3] As a rule the calorimeter content is stirred in order to establish thermal equilibrium. Only the superimposition of the cooling power w_A and the constant power of stirring w_0 (3.3) can then be observed.

$$C\frac{dT}{dt} = -\kappa C(T - \Theta) + w_0 \tag{9.1b}$$

The constant power of stirring can be eliminated by measuring the temperature rise dT/dt or $\Delta T/\Delta t$ for two dissimilar differences in temperature $(T - \Theta)$:

$$\kappa = \frac{(\Delta T/\Delta t)_{\text{II}} - (\Delta T/\Delta t)_{\text{I}}}{\overline{T}_{\text{I}} - \overline{T}_{\text{II}}} \tag{9.1c}$$

The constant temperature Θ of the enclosure is eliminated at the same time; $\overline{T}_{\text{I}}$ and $\overline{T}_{\text{II}}$ are the mean values found during the time of observation.

Another method for determining κ and w_0 was discussed in Section 3.1.2. The previous discussion presupposed a thermal equilibrium between the calorimeter vessel and the enclosure. During a thermometric titration, however, the addition of the titrant disturbs this equilibrium continually and this transition must be taken into account. The problem can be tackled by using the heat-balance equation

$$\nabla^2 T(x,t) = \frac{\rho c}{\lambda} \frac{\partial T(x,t)}{\partial t} \tag{9.7}$$

In (9.7), ρ is the density, c the specific heat, and λ the thermal conductivity of a heat-conducting body; ∇^2 is the Laplace operator.

Transients must be taken into account in the design of a calorimeter, since the uncertainty in heat leak correction should be made negligible in precision calorimeters and the true temperature of the calorimeter should be measured without time lag. The difficulties arising in this connection are essentially caused by the mechanical link between the calorimeter vessel and its lid. Insulating material, for instance, glass, is generally used for the support. The following discussion shows how a thermal equilibrium between calorimeter content and the lid arises in the calorimeter after a temperature rise ΔT of the solution. The heat is conducted by the support of the calorimeter (the glass tube in Fig. 9.1), which is fixed to the lid of the enclosure with the help of silver conductive epoxy.

The temperature of the point of fixing is exactly that of the enclosure (Θ), and that of the lower sectional area is identical to that of the calorimeter content (T). An adequate model for this problem of heat transfer is given in Fig. 9.2.

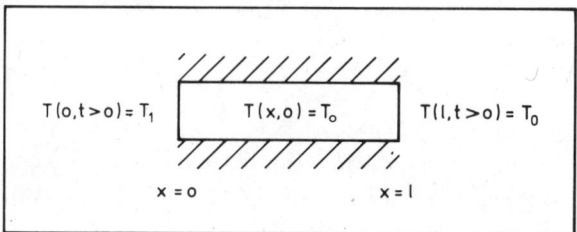

Fig. 9.2. Heat transfer in a bar with insulated lateral surface.

The laterally completely insulated conductor (Pyrex tube) has the uniform temperature $T_0 = T(x,0) = \Theta$ at the time $t=0$. At $t>0$ there is a temperature rise to $\Theta + \Delta T = T_1$ at the surface $x=0$. The temperature distribution at $t>0$ in the conductor can be described as the solution of the differential equation (9.7) by taking into account the initial and limiting conditions.[4]

$$T(x,t) = \Theta + \Delta T - \frac{\Delta T}{l} x - \frac{2}{\pi} \sum_{n=1}^{\infty} \frac{\Delta T}{n} \exp\left(-\frac{n^2 \pi^2 \lambda t}{\rho c l^2}\right) \sin\left(\frac{n\pi x}{l}\right) \quad (9.8)$$

The temperature $T(x,t)$ at the point x and at time t deviates from the equilibrium temperature $T(x,\infty) = \Theta + \Delta T - (\Delta T / l) x$ by the amount

$$\epsilon(x,t) = T(x,\infty) - T(x,t) \quad (9.9)$$

The dependence on time of the heat transfer coefficient, which has been measured experimentally,[5] and the dependence on time of the heat capacity, which has also been observed, can be explained in this way.

PRINCIPLES OF DESIGN

The example chosen uses the following data for Pyrex: $\lambda = 0.0027$ cal cm^{-1} sec^{-1} K^{-1}; $c = 0.18$ cal g^{-1} K^{-1}; $\rho = 2.23$ g cm^{-3}. The temperature function $T(x,t)$ was calculated for $\Delta T = 0.1$ K at the points $x/l = 0.25$, 0.5, and 0.75 at the times $t = 10$, 100, 500, and 1000 sec. Table 9.3 contains the results in the form $T(x,t) - \Theta$. The corresponding equilibrium temperature $T(x, \infty) - \Theta$ is also given for comparison. The arithmetic mean was taken for the deviation $\epsilon(t)$ from this equilibrium temperature between $x = 0$ and $x = 1$.

The product of this mean value $\bar{\epsilon}(t)$ with the heat capacity C^* of the calorimeter support represents the deviation at time t of the heat taken up by the calorimeter support from the equilibrium value ($t = \infty$). The quotient $C^*\bar{\epsilon}(t)/(C\Delta T)$, where $C\Delta T$ is the complete heat effect, is the relative error in determining the heat effect caused at the time t by the time-dependent equilibrium establishment.

It is evident from Table 9.3 that this error is not negligible under 1000 sec ($<0.01\%$). Sufficiently exact measurements can be obtained, however, at $t > 100$ sec (0.1%). A fast equilibrium establishment cannot be achieved with a glass calorimeter support. It is, however, possible to minimize the resulting errors sufficiently by making the heat capacity C^* of the support small. The heat capacity in our example is 1.8 cal K^{-1}, whereas the total heat capacity C of the calorimeter (filled with 200 ml water) is about 220 cal K^{-1}. By using glass tubes with thin walls (e.g., thickness 0.07 cm instead of 0.2 cm) the errors listed in Table 9.3 could be reduced to one third.

Our discussion shows that it is possible to estimate the dynamic properties of a calorimeter with uncomplicated methods. Such calculations produce results that otherwise could only be achieved by elaborate experiments.[5] Section 9.2 discusses the application to calorimeter design of the insights gained. In the present context we wish to discuss only the method for determination of the equilibrium time.[5] It is based on (9.1c), which gives Newton's cooling coefficient κ for a system in thermal equilibrium. The temperature rise $(\Delta T/\Delta t)_\text{I}$ is that of the system in thermal equilibrium, and $(\Delta T/\Delta t)_{\text{II},t}$ is the rise for different time intervals after the thermal equilibrium has been disturbed. We then get

$$\kappa' = \frac{(\Delta T/\Delta t)_{\text{II},t} - (\Delta T/\Delta t)_\text{I}}{\bar{T}_\text{I} - \bar{T}_\text{II}}$$

$$\lim_{t \to \infty} \kappa' = \kappa \qquad (9.10)$$

Thermal equilibrium is reached if (9.10) gives a constant value for κ'. The time up to attainment of equilibrium is called the *equilibrium time*.

TABLE 9.3. Establishment of Temperature Equilibrium in the Calorimeter Support

	$T(x,\infty)-\Theta$	$T(x,t)-\Theta$	$\epsilon(t)$	$\bar{\epsilon}(t)$	$C^*\bar{\epsilon}(t)/C\Delta T$ ($^0/_{00}$)
$t=10\sec \Delta T=0.1\,K$					
0.00	0.100	0.100	0.000		
0.25	0.075	0.004	0.071		
0.50	0.050	0.000	0.050	0.037	30
0.75	0.025	0.000	0.025		
1.00	0.000	0.000	0.000		
$t=100\sec \Delta T=0.1\,K$					
0.00	0.100	0.100	0.000		
0.25	0.075	0.052	0.023		
0.50	0.050	0.019	0.031	0.019	1.5
0.75	0.025	0.005	0.020		
1.00	0.000	0.000	0.000		
$t=500\sec \Delta T=0.1\,K$					
0.00	0.100	0.100	0.000		
0.25	0.075	0.074	0.001		
0.50	0.050	0.048	0.002	0.001	0.07
0.75	0.025	0.024	0.001		
1.00	0.000	0.000	0.000		
$t=1000\sec \Delta T=0.1\,K$					
0.00	0.100	0.100	0.000		
0.25	0.075	0.075	0.000		
0.50	0.050	0.050	0.000	0.000	0.00
0.75	0.025	0.025	0.000		
1.00	0.000	0.000	0.000		

Carefully designed calorimeters[5,6] possess an equilibrium time ranging from a few seconds up to 2 or 3 min; that of traditional Dewar calorimeters is of the order of 60 min.

The aspects discussed so far are valid in their entirety for the constant-temperature environment calorimeter, the type most frequently used in thermometric titration. The characteristic feature of this type of calorimeter is that the temperature of the enclosure is constant and different from the changing temperature of the calorimeter vessel. The aim is to have the best possible thermal insulation of the calorimeter vessel from the enclosure. The above discussions and calculations have shown that strictly adiabatic behavior cannot be achieved.

The adiabatic calorimeter is characterized by the calorimeter vessel having the same temperature as its environment. The temperature of the enclosure is adapted to that of the calorimeter vessel with the help of a temperature regulator. Since $T-\Theta \approx 0$, heat transfer is negligibly small. A quick change of temperature will bring about a lag between the temperature of the calorimeter vessel and that of its enclosure. During this time lag the calorimeter is not strictly adiabatic. Because of its inertia and its elaborate design, this type of calorimeter has not been used in thermometric titration.

Another possibility of rendering negligibly small the undesirable heat transfer between the calorimeter vessel and its enclosure consists in maintaining the temperature of the vessel constant and equal to that of the enclosure Θ (constant-temperature calorimeter). The power of the reaction is compensated electrically,[7-10] in exothermic reactions with the help of the Peltier effect and in endothermic reactions by electrical heating. As the temperature of both calorimeter vessel and enclosure is constant, thermal equilibrium is established quickly. This type of calorimeter is now being used in titration calorimetry.

9.1.2 TEMPERATURE MEASUREMENT IN TITRATION CALORIMETRY

The most important qualities demanded of a temperature sensor are

1. High sensitivity so that small temperature deviations may be detected.
2. Rapid thermal response so that the temperature sensor is in thermal equilibrium with the calorimeter and its content.
3. Stability, at least within the measurement period.

Other conditions influencing the selection of the most suitable temperature sensor are size, thermal qualities, and the furnishings necessary for its working.

Liquid-in-glass thermometers, which were used formerly, have sufficient absolute accuracy for precision measurements (about 0.01 K), but their temperature resolution (about 0.001 K) is inadequate in many cases. Other disadvantages are their size and the impossibility of recording the temperature. The problem is solved by using thermocouples, multijunction thermocouples, metal resistance thermometers, and thermistors. The introduction of thermistors was one of the most important steps in the development of thermometric titration, but the use of other temperature sensors can be of advantage for special problems.

Thermocouples and Multijunction Thermocouples. A thermocouple consists of two different conductors, A and B, which are linked as shown in Fig.

9.3a. The temperature-dependent potentials $E_{AB}(T)$ of the junctions yield a potential difference $E = E_{AB}(T_1) - E_{AB}(T_2) = f(T_1 - T_2)$ if there is a temperature difference $T_1 - T_2$. When the temperature of one of the junctions is kept constant, for instance, the temperature T_2 at the junction (2), the emf E will be a function of the temperature T_1 of the junction (1) only. The thermocouple can be used to measure both temperature differences and absolute values. To measure absolute temperatures, the function $E = f(T_1 - T_2)$ must be known. Melting ice is used commonly for the constant temperature T_2. To measure the emf of the thermocouple, the circuit must be opened and connected to an instrument. In Fig. 9.3b the conductor B is interrupted and connected to a third conductor C, which is the link with the instrument. Thus two more junctions (3) and (4) are created, the potentials of which are the same and, therefore, cancel out, provided they have the same temperature. Basically any pair of conductors can be called a thermocouple, but only a few combinations have become established in practice. The criteria of usefulness are

1. The temperature sensitivity dE/dT of their emf.
2. Homogeneity, and the mechanical and chemical stability of the wire used as a conductor.
3. Electrical and thermal conductivity.

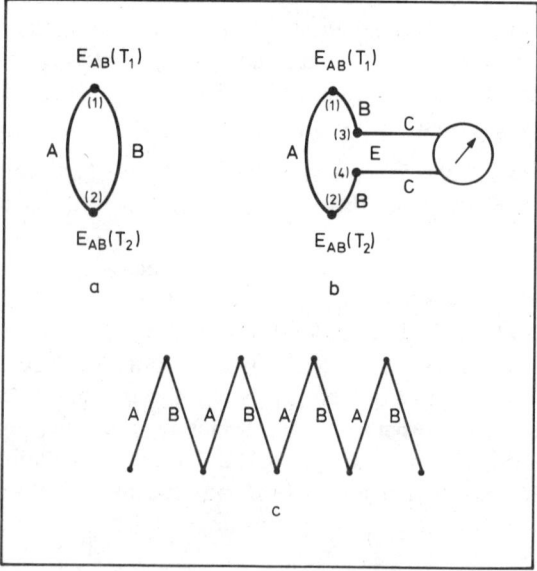

Fig. 9.3. Sketch of two conductors linked to form a thermocouple (*a*), a thermocouple with a measuring instrument (*b*), and a multijunction thermocouple (*c*).

Some properties of thermocouples most frequently used today, designated as types S, R, J, T, K, and E by the Instrument Society of America (ISA), are given in Table 9.4.

TABLE 9.4. Properties of Commonly used Thermocouples

ISA Code	Material	Sensitivity at 25°C μV K^{-1}	Working Range °C		
S	90%Pt–10%Rh/Pt	5.5	0	to	1100
R	87%Pt–13%Rh/Pt	5.5	0	to	1400
J	Fe/constantan	55	−190	to	760
T	Cu/constantan	40	−200	to	400
K	Chromel/alumel	40	−190	to	1200
E	Chromel/constantan	60		to	900

The emf's of thermocouples connected in series are additive (Fig. 9.3c). Multijunction thermocouples or thermopiles, as such combinations are called, possess a high temperature-sensitivity. They can be of special advantage in measuring small temperature differences. Measuring problems of this kind arise in twin calorimetry, continuous flow, or adiabatic calorimetry. The thermocouples are arranged so that the wires between the points of temperature measurement are situated in a zone of constant temperature. Spurious emf's, caused by the inhomogeneities of the wires, are thereby avoided.

Electrical Resistance Thermometers. The resistance R of a metal increases with increasing temperature. The relative coefficient $\alpha = (1/R)(dR/dT)$ is temperature independent at a first approximation. It is lowered by impurities, so that only highly purified metals are used to construct resistance thermometers. The most frequently used are platinum ($\alpha = 3.92 \cdot 10^{-3}$), nickel ($\alpha = 6.8 \cdot 10^{-3}$), and copper ($\alpha = 4.3 \cdot 10^{-3}$); platinum thermometers are favored because of their high stability.

The resistance of commercial platinum sensors ranges from 25.5 to 500 Ω (at 0°C). The small absolute temperature coefficient $\alpha R = 0.1$ Ω K^{-1} of 25.5-Ω thermometers (e.g., standard thermometer 8163-B, Leeds and Northrup Co.) makes it necessary to use special resistance bridges (Mueller bridge, Smith bridge). Both dimension and time constant (about 16 sec) render the use of this type inadvisable for thermometric titration.

Miniature sensors (e.g., type S31-A, Minco Products) with a time constant of 1.3 sec and $R=470$ Ω at 0°C are more suitable. They have $\alpha R \approx 1.8$ Ω K^{-1}; this higher temperature coefficient lowers the demands to be made on the measuring bridge. The stability of miniature sensors over their whole operating range is less than that of standard sensors, 0.03K against 0.001 K. As they are used only for a very limited temperature range, a stability of 0.001 K in 6 months can be relied upon; it has been sufficiently tested experimentally.[11]

A criterion for the quality of a platinum resistance thermometer is the resistance ratio, R_{100}/R_0 (resistances at 100°C and 0°C, respectively); a minimum value of 1.3920 is required for standard thermometers, and this value is reached by miniature sensors also. Resistance thermometers are calibrated with the help of fixed points of temperature. Platinum resistance thermometers are predominantly used for absolute measurements; small differences of temperature are more easily measured with thermistors.

The measurement of a temperature change of $\Delta T = 10^{-4}$ K with the help of a platinum sensor demands measurement of a relative change of resistance of $4 \cdot 10^{-7}$. The use of manganin resistors with a temperature gradient of $\pm 10^{-5}$ K^{-1} means that the measuring bridge must be thermostated to ± 0.02 K.

Thermistors. Thermistors are semiconductors, which are composed of metal oxides, such as of manganese, nickel, cobalt, copper, iron, titanium, and uranium. They are available over a very wide range of resistances and in a large variety of forms, such as beads, disks, rods, glass probes, and so forth. Thermistors are defect semiconductors, the conduction band of which is separated from the valence band by an energy gap ΔE of the order of kT. The distribution of electrons between valence and conduction band is governed by a Boltzmann-type distribution and the resistance temperature relation may, therefore, be approximated by the exponential function

$$R(T) = Ae^{B/T} \tag{9.11}$$

or, more accurately,[12] by the empirical formula

$$R(T) = Ae^{B/(T+\Theta)} \tag{9.12}$$

A typical value for B is 3500 K. The temperature coefficient

$$\alpha = \frac{1}{R}\frac{dR}{dT} = -\frac{B}{T^2}$$

has the large negative value of $-4 \cdot 10^{-2}$ K^{-1}. The large temperature coefficient and the high value of the resistance permit small temperature changes to be measured (without compensation of the lead resistances) by means of simple Wheatstone circuits.

In 1959 Schlyter[13] could still state: "Actually R_{25} drifted appreciably with time, whereas the temperature coefficient C_t kept constant. Moreover, the thermistor resistances sometimes altered unpredictably (thermistor 'jumps'), which completely spoiled the precision of temperature-difference observations...". Since then the stability of thermistors has been improved, especially by artificial ageing and special compounding of the metal oxides. The design, too, contributes a considerable share to the improved stability (bead thermistor, glass-probe thermistor). Glass-probe thermistors exposed to a temperature of 100°C are stable to 0.1% per year; in long-time testing at 25°C even to 0.02% per year.[14] Jordan[15] states that the parameters of a thermistor change less than 0.1% during a period of 6 months when a current of 0.1 to 1 mA is passed unceasingly through it. Since their first use by Linde, Rogers, and Hume,[16] thermistors have gained considerable ground in thermometric titration. If the measuring bridge is designed carefully and if low-noise and low-drift amplifiers are used, temperature changes of about 10^{-5} K can be measured. It is not reasonable to aim at a higher resolution in temperature measurement, as even efficient stirring yields temperature fluctuations of this order of magnitude in the calorimeter vessel.

At the present stage of technological development, thermistors are the most useful temperature sensors for thermometric titration. They have a high sensitivity, rapid thermal response (~0.3 sec for bead thermistors), low heat transfer coefficient, low heat dissipation, and sufficient stability. A recently developed positive temperature coefficient thermistor with a sensitivity of 0.2 K^{-1} offers new opportunities for measuring temperature.[17] Little is known about the stability of this device, however.

Electrical Equipment for Temperature Measurement in Titration Calorimetry. The choice of instrument for emf measurements with thermocouples depends on the magnitude of the emf's and the purpose for which the thermocouples are to be used. In twin calorimetry, for example, multijunction thermocouples are used with a sensitive instrument such as a galvanometer or a convenient dc amplifier. This simple method will always be used for the measurement of a relatively small temperature difference. Larger temperature differences or the measurement of absolute temperature require potentiometric compensation of the emf.

Temperature measurements with the help of resistance sensors are conducted with bridge or potentiometric methods. Since the electrical qualities of the sensors differ greatly, special measuring bridges have had to be developed. The resistance of Pt-sensors rarely exceeds 500 Ω; thermistors have, as a rule, a resistance between 1000 Ω and 100 KΩ.

The relatively low resistance and small temperature coefficient of platinum sensors makes an exact compensation of the lead resistance imperative. Platinum sensors for precision measurements must have three or four leads that enable the real sensor resistance to be measured in combination with a Mueller (Fig. 9.4a) or a Smith bridge. The lead resistance can be eliminated completely only with the help of potentiometric methods (Fig. 9.4b).

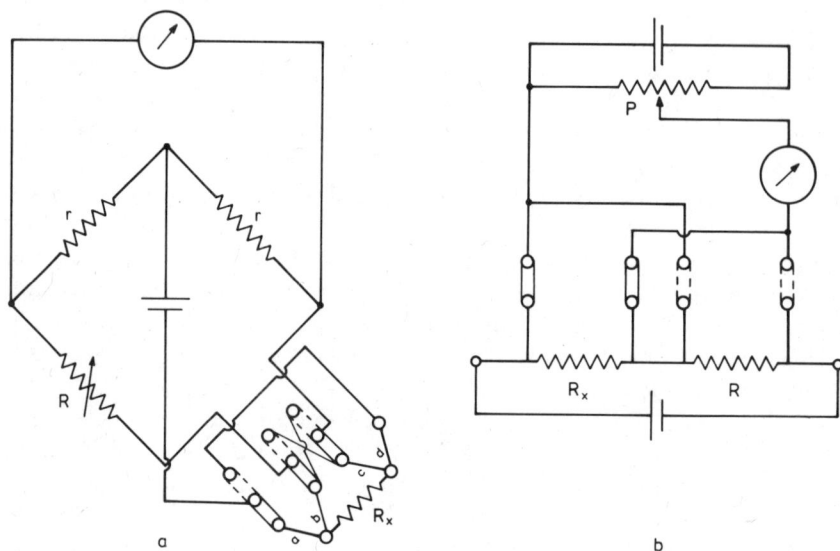

Fig. 9.4. Schematic drawing of (a) a Mueller bridge, and (b) a potentiometer circuit for the determining of the resistance of a platinum thermometer. (a): R_x, sensor resistance; a,b,c,d, resistances of the four leads connected with a reversing commutator; R, adjustable resistance; and r, resistances of the ratio arms. (b): R_x, sensor resistance; R, standard resistance; and P, potentiometer.

If high-ohmic, sensitive thermistors are used for temperature measurements, compensation of the lead resistance is unnecessary and a simple Wheatstone bridge may be used (Fig. 9.5). Temperatures are usually measured by balancing the measuring bridge, but titration calorimetry records the unbalance voltage of the bridge as a continuous function of time (or the volume of titrant added).

If the bridge (Fig. 9.5) is out of balance, the output voltage (unbalance voltage) is

$$E_0 = E_s \left[\frac{R_1}{R_1 + R_2} - \frac{R_3}{R_3 + R(T)} \right] \qquad (9.13)$$

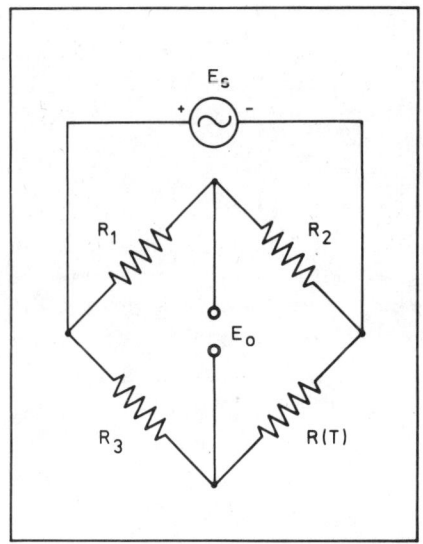

Fig. 9.5. Schematic drawing of a thermistor bridge circuit.

If the resistors R_1, R_2, R_3, and the input voltage E_s are constant, we get

$$\Delta E_0 = E_s \left\{ \frac{R_3}{[R_3 + R(T)]^2} \right\} \Delta R \qquad (9.14)$$

From (9.11) we can derive

$$\Delta R = -R \frac{B}{T^2} \Delta T = -\frac{B}{T^2} A e^{B/T} \Delta T \qquad (9.15)$$

The bridge is taken to be in equilibrium at the temperature T_0, then $R(T_0) = R_2$ if $R_1 = R_3$.

For a temperature

$$T = T_0 \left(1 + \frac{\Delta T}{T_0} \right) \qquad (9.16)$$

differing by ΔT from the equilibrium temperature T_0, we obtain a change of resistance ΔR, from (9.15) by expanding in series and truncating after the linear term:

$$\Delta R = -R(T_0) \frac{B}{T_0^2} \left[1 - \left(\frac{B}{T_0^2} + \frac{2}{T_0} \right) \Delta T + \cdots \right] \Delta T \qquad (9.17)$$

The change of the resistance ΔR is, even at a first approximation, not a linear function of ΔT. The deviation from linearity is about -0.3% for a temperature change $\Delta T = 0.1$ K at $T = 300$ K. Further, it must be noted that the unbalance voltage E_0 of the Wheatstone bridge does not itself change linearly with $R(T)$.

According to (9.14) and (9.15), the unbalance voltage is a function of the temperature deviation ΔT. Tyson, McCurdy, and Bricker[18] have shown that, at suitable R_3 values, ΔE_0 can be made to vary linearly with the temperature change ΔT when this is near the reference temperature T_0. For this it is necessary to linearize the change in the voltage drop of the thermistor itself; the following condition must then apply:

$$\frac{d^2}{dT^2} \frac{R_3}{R_3 + R(T)} = 0 \qquad (9.18)$$

By using (9.11), we get

$$R_3 = R(T_0) \frac{B - 2T}{B + 2T} \qquad (9.19)$$

For a temperature range of ± 4 K around the reference temperature of $T_0 = 297.2$ K, the deviation from linearity is about 1%; for a range of ± 1.5 K it is only 0.1%.

Another possibility of making ΔE_0 linearly dependent on ΔT consists in rendering the temperature function of the thermistor equation (9.17) linear and, at the same time, keeping the current flowing through the thermistor constant. This linearizes the voltage drop at the thermistor.

The temperature function of the thermistor resistance can be linearized in several ways, in all of which the thermistor resistance $R(T)$ is connected in series, in parallel or in a combination of both ways with suitable fixed resistors. The most frequent combinations are shown in Fig. 9.6. The circuit arrangement Fig. 9.6a will always suffice for the purposes of thermometric titration. Nordon and Bainbridge have shown[19] that a resistance

$$R_p = R(T_0) \frac{B - 2T}{B + 2T} \qquad (9.20)$$

is required. Over a range of 50 K, the deviation from linearity is 0.5% and, for a range of 18 K, only 0.05%. The layout of Fig. 9.6b and c permits linearization over an even larger range.[20]

When the exact value of the reaction enthalpy ΔH_R is sought, the problem of linearization of the measuring bridge can be bypassed. The instrument used to measure the temperature is calibrated by feeding into

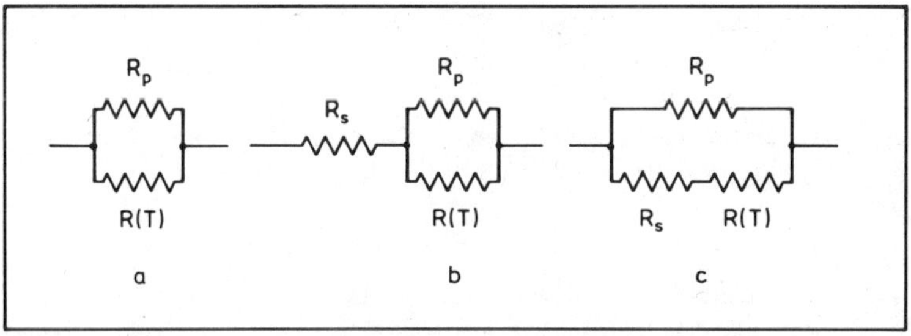

Fig. 9.6. Circuits for the linearization of the thermistor resistance.

the calorimeter definite quantities of electrical heat ($W_{el}\Delta t$). We get

$$\Delta H_R = \frac{W_{el}\Delta t}{\Delta X_{el}} \Delta X_R \qquad (9.21)$$

ΔX (scale units) is a measure of the unbalance of the bridge. If $\Delta H_R \approx W_{el}\Delta t$, then $\Delta X_R \approx \Delta X_{el}$; the nonlinearity of the measuring bridge is thus not relevant.

Finally, attention should be drawn to the fact that methods aiming only at indication of the end point can dispense with linearity. In such cases it is, however, no longer possible to draw conclusions about the reaction process from the shape of the titration curve (see Chapter 2, Section 2.3.6).

Varying attention has been paid in publications on thermometric titration to the problem of linearizing the unbalance voltage. Jordan and Alleman[21] state that their measuring bridge is linear to ±1% in the range of ±1 K. This is sufficient only when small changes of temperature (about 0.5 K) are to be measured. With larger changes of temperature the growing deviation from linearity becomes a disturbing factor. Everson[22] describes a linear thermistor bridge that permits direct reading of temperature over a range of 18 to 30°C. Temperature measurement with the help of thermistors in differential calorimeters requires an exact linearity, and the sensitivities of two thermistor sensors must be adapted to each other;[18] otherwise the temperature changes common to the two calorimeters do not cancel out. A similar measuring problem arises in flow calorimetry: The difference must be measured between the mean temperatures of two solutions entering the reaction chamber and the temperature of the solution leaving it. To achieve this, three thermistors must be linearized and adapted.[23]

It is often an advantage to record the unbalance voltage of the measuring bridge or the emf of the thermopile. If the unbalance voltage is recorded directly, a sensitivity of 1 mV over the whole range of the recorder makes it possible to record a voltage change of 1 μV. A bridge with two identical branches $[R_1 = R_2 = R_3 \approx R(T)]$ and a supply voltage of 1.35 V (Mallory cell), using a thermistor with $B = 3500$ K, will have a temperature sensitivity of the unbalance voltage $\Delta E_0/\Delta T \approx 13 \mu V/10^{-3}$ K. The recorder sensitivity of 1 μV permits a temperature change of 10^{-4} K to be registered. The same sensitivity is found in a thermopile consisting of 30 copper–constantan thermocouples.

Higher sensitivity can be attained with the help of suitable dc amplifiers (e.g., Keithley model 150B). The limits to temperature resolution set by the inhomogeneous temperature distribution in the calorimeter (10^{-5} K) have been pointed out already.

As recording instruments, either x-t recorders or x-y recorders are used, the latter permitting a control via the burette feeding in the titrant. It is an improvement to replace the analogous recorder by a digital recorder (e.g., printer) unless the evaluation of the thermogram is to be based on the strip chart itself. The digital method is useful, too, for evaluating the thermodynamic data of the reaction with the help of a computer.

Both types of recording can be used for titrations with continuous or discontinuous addition of the titrant; digital recording is more advantageous when the titrant is added discontinuously. The advantages of a new and automatic method with digital recording and discontinuous addition of the titrant are discussed in Section 9.2.1.

9.1.3 THE ADDITION OF TITRANT SOLUTION

In solution calorimetry the "titrant" is commonly sealed into a glass ampul and the ampul is immersed in the solution in the calorimeter vessel. After temperature equilibrium has been reached, the ampul is broken.[5] The advantage of this method is that there is no heat of mixing and it is possible to add the titrant in the solid state or as a solution. The disadvantage is that a titration diagram can be drawn only after a number of independent measurements with varying ratios of the reactants have been made. A variant of this method consists in feeding the titrant first into a suitable burette within the calorimeter vessel and mixing the titrant with the content of the calorimeter after temperature equilibrium has been established.[24] This method permits the titrant to be added in several discontinuous steps, but it is very time-consuming.

Dutoit and Grobet[25] describe thermometric titration as a volumetric method and hence use a burette for adding the titrant. Many variations of this principle have been applied up to the present time. Difficulties were created in the early days of thermometric titration by the uncontrolled temperature of the titrant. A thermostat jacket for the burette represents a first way of obtaining a controlled temperature. This can be done with normal burettes and the motor-driven syringe burettes used for the continuous addition of titrant. It still leaves the question of the temperature of the solution in the tip of the burette, which is not thermostated. The error arising can be minimized by thermal insulation of the tip and by making it as small as possible.[26] Schlyter[13] solved the problem by first feeding the titrant from the burette into a bulb of 50 ml that is thermostated together with the calorimeter vessel. From this chamber the titrant is conducted via a capillary tube into the calorimeter vessel. A variant of Schlyter's method used for continuous addition can be found in Danielsson, Nelander, Sunner, and Wadsö,[27] who feed the titrant solution into the calorimeter vessel via a heat exchanger. Christensen, Izatt, and Hansen[6] used a storage container for the titrant solution immersed in the thermostat liquid. The solution is expelled with the help of a burette filled with mercury and flows through a thin tube, which is also immersed in the thermostat liquid, into the calorimeter.

Unavoidably complex devices for thermostating the titrant are necessary if thermodynamic titration data are sought from a thermogram. If, for instance, the total amount of the titrant is 10% of the volume of the sample, a deviation of $\pm 10^{-2}$ K of the temperature of the titrant results in a temperature error of $\pm 10^{-3}$ K during the reaction period.

Normal burettes show a deviation from standard volumes of as little as 0.1%. Motor-driven burettes, originally developed by Lingane[28] and suggested by Linde, Rogers, and Hume[16] for use in thermometric titration, can be designed with about equal accuracy.

Figure 9.7 shows a syringe burette used in the author's laboratory. It can be driven manually as well as by a synchronous or stepper motor. The plunger of a gas-tight syringe (Hamilton, Type 1010LL), surrounded by a thermostat jacket, is moved by a micrometer screw. A dial gauge (resolution 0.002 mm) shows the position of the plunger.

To calibrate the burette, the plunger is driven down successive 5-mm distances and the weight of the volumes of water expelled are determined. Inaccuracies inherent in the system, mainly due to irregular dimensions of the syringe, limit the absolute accuracy of the syringe burette to about 0.05%.

9.1.4 CALIBRATION OF TITRATION CALORIMETERS

The basic equation of thermometric titration can be used only if the heat capacity $C(t)$ of the calorimeter is known. Cases in which the determination of the heat capacity were part of the method of evaluation were discussed in Section 3.1.2. Instrumental design of the different methods of calibration is discussed in the following paragraphs.

The determination of the heat capacity $C(t)$ is based on either electrical calibration or calibration with the help of a standard reaction.

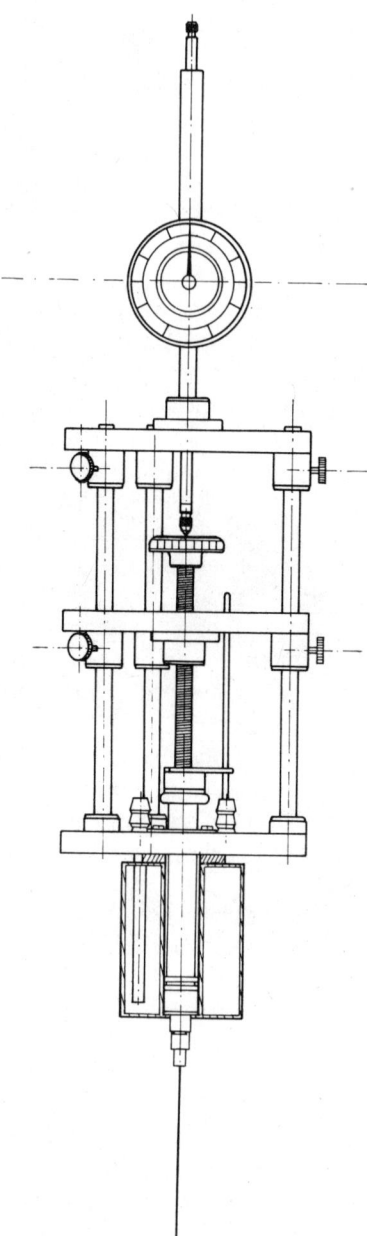

Fig. 9.7. Thermostated syringe burette.

Electrical calibration. For electrical calibration (see Section 3.1.2), a definite amount of heat is fed into the calorimeter from a calibration heater. A simple circuit for this is shown in Fig. 9.8. It consists of the calibration heater R_H in series with a standard resistor R_S and a dc power source, such as a lead storage cell or a precision dc power supply.

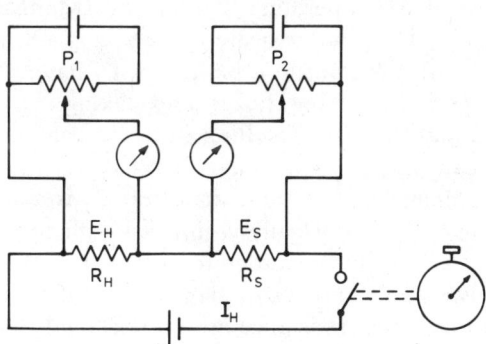

Fig. 9.8. Circuit for electrical calibration.

The voltage drop E_S across the standard resistor R_S is measured by a potentiometer P_2 giving the current I_H in the heating element by the relationship $I_H = E_S/R_S$. A similar potentiometer P_1 measures the voltage drop E_H across the heating element. The power dissipated in the heater is

$$W = E_H I_H = E_H \frac{E_S}{R_S} \tag{9.22}$$

The electrical power entering the calorimeter is Wt, t being the duration of the current flow through the heater. The time can be simply and efficiently measured by means of an electric stopwatch synchronized with the heating current.

After the heating current switching on, the temperature of the heater increases until thermal equilibrium is reached. The temperature rise ΔT is defined by the relationship $\Delta T = aW$, in which a is the thermal resistance (K W^{-1}) between the heater and its environment (calorimeter). A thermal resistance of 20 K W^{-1} and a heating power of 0.5 W results in a $\Delta T = 10$ K. During the few seconds between switching on the current and the establishment of thermal equilibrium, the heating current, along with the heating power, is changing exponentially. Theoretically, it would be possible to determine E_H and E_S as functions of t during this time of transition and to define the electrical power during the complete heating period as

$$\frac{1}{R_S} \int_{t_0}^{t} E_H(t) E_S(t) \, dt \qquad (9.23)$$

It is, however, better to use methods in which the influence of the transition period is rendered negligibly small. This can be done by using resistance wire of low-temperature coefficient (Manganin, Evanohm, Karma, all ± 10 ppm K^{-1}) for the heater, and having a heating period that is long compared with the transition period, or by having a heater with a short transition period.[29] Using (9.22) and taking E_H and E_S to be constant, we find that the error resulting from the change in resistance of the heating wire can then be kept to less than 0.01%.

A correct determination of the potential drop E_H is more difficult. Both ends of the heating wire are usually connected with potential leads. The wires are as thin as possible in order to reduce heat conductance. The current leads, however, cannot be designed according to this principle, since thermal and electrical resistance are proportional to each other.

Reducing the cross-section of the lead increases the electrical resistance, which leads to a rise of temperature due to joule heat. The best cross-section is a compromise, depending on the resistance of the heater: the larger the resistance of the heater, the smaller the heating current for a constant heating power and, therefore, the smaller the rise in temperature of the leads. The resistance of the heater should not be less than 10 Ω; a value of 100 Ω is more suitable.

The heater circuit can be elaborated or simplified, depending on the accuracy to be achieved. Measurement of the time, in particular, can be improved by using electronic watches. Complicated potentiometric measurement of voltage can be circumvented by using high-impedance digital voltmeters. The use of printers allows automatic recording at constant time intervals.

The substitution of lead cells by an electronically controlled precision voltage source offers two important advantages:

1. The voltage becomes constant to 0.01% after a short heating period (30 min). The output resistance is so small that the voltage is largely independent of the load. The use of a dummy resistance can thus be avoided.
2. The voltage, and hence the heating power, can be changed continuously.

The amount of electrical heat can be determined with high accuracy (0.01%). Nevertheless, it is advisable to use chemical reactions of exactly known heats of reaction for the calibration of calorimeters also. Electrical calibration should be checked by an independent method, since systematic errors do not appear in the standard deviation.

Calibration by a standard reaction. Standard reactions in general use usually serve for calibration. Thus conductivity cells, for instance, are calibrated exclusively by using exactly defined KCl solutions. There are no such standard methods for the calibration of titration and solution calorimeters, but a number of methods has been proposed already. The one most frequently used makes use of the heats of reaction of the following acid-base reactions:

1. The heat of neutralization.
2. The heat of reaction of tris(hydroxymethyl)aminomethane with 0.1 M hydrochloric acid.

A comprehensive treatment of standard reactions can be found in Section 4.1.3, where the special advantages of the reaction mentioned under 2. are discussed.

If both electrical and chemical calibration agree within their limits of error, then systematic errors can be excluded. The standard deviations, when only one of these methods is used, permit statements about the reproducibility, whereas systematic errors may be underestimated or completely overlooked.

9.2 CALORIMETRIC EQUIPMENT FOR SPECIAL PURPOSES

Thermometric apparatus should satisfy two demands: first, the analytical search for the end point; and, secondly, the calorimetric inquiry that aims at the most complete determination possible of the thermodynamic data of a titration reaction. The two demands do not always carry equal weight, a fact that has led to the development of special equipment.

9.2.1 TITRATION CALORIMETERS FOR ACCURATE ENTHALPY CHANGE MEASUREMENT

Even carefully designed titration calorimeters using commercial Dewar flasks as reaction vessels do not enable the heat of reaction to be determined with higher accuracy than 0.5%. This is due to the high equilibrium time (about 60 min) and the ill-defined boundary with the surroundings. Their simplicity is an advantage and their accuracy is frequently adequate. For a long time more exact measurements could be achieved only by solution calorimetry, the disadvantage of which was the time-consuming determination of one measuring point after another. The final goal of combining the thermometric method, providing all the data of the reaction in a single experiment, with the accuracy of solution calorimetry was approached only after a long time. In recent years there has been a systematic development of precision calorimeters for titrations.

Sunner and Wadsö[5] have studied the properties of various models of vacuum jacket calorimeter of the constant-temperature environment type. The best design proved to be a thin-walled metal container as calorimeter vessel fixed to the lid by means of a thin-walled glass tube. The calorimeter designed by Sunner and Wadsö, originally developed as a solution calorimeter, satisfies the conditions listed in Section 9.1: a low heat capacity of the support; exact boundary between the calorimeter vessel and its environment, resulting in an equilibrium time of 2 min; evacuated space between the calorimeter vessel and the jacket; a thermostated bath with a constant temperature Θ enclosing the calorimeter. The LKB titration calorimeter Type 8721-2 (Fig. 9.9) is a further development of their calorimeter for titration purposes.

The exchangeable calorimeter vessel with a content of 25 or 100 ml is made of thin-walled Pyrex and contains a 2000-Ω thermistor, a 50-Ω calibrating heater, and a stirrer. The titrant is added with the help of a motor-driven burette through a heat exchanger and is fed into the calorimeter vessel through a capillary tube. The calorimeter vessel is surrounded by a chromium-plated brass container that can be evacuated. This container is immersed in the thermostat liquid maintained at a temperature constant to 10^{-3} K. The electronic equipment consists mainly of a system for temperature measurement and an electrical calibrating system. The temperature is measured by the thermistor by balancing the measuring bridge or by registering with a recorder. The calibrating system permits a step-by-step setting of the heating power and the presetting of definite calibrating times (1 to 990 sec) with a high degree of accuracy (± 0.003 sec) and reproducibility (± 0.003 sec).

Christensen, Izatt, and Hansen[6] have shown that the disadvantages of Dewar vessel calorimeters can be avoided by proper design (Fig. 9.10). The inner vessel is a thin-walled (0.6-mm) flask; the narrowing of the cross-section at the neck of the flask defines the boundary of the calorimeter vessel. The heat transfer is low, $\kappa = 2 \cdot 10^{-3}$ min^{-1} and the equilibrium time is only 1 to 3 sec. In order to profit from this favorable equilibrium time, sample and titrant must be mixed very quickly and a thermistor with a low time constant τ used. Figure 9.11 shows the calorimeter insert designed according to these principles with a thermistor ($\tau = 0.3$ sec), heater, and titrant delivery tip.

Special attention was paid to the constant temperature of the environment and the titrant. The calorimeter is immersed in the inner bath of a twin thermostat of temperature constant to $\pm 3 \cdot 10^{-4}$ K over a period of 24 hr. The titrant is stored in the same bath, and the solution is conducted to

CALORIMETRIC EQUIPMENT FOR SPECIAL PURPOSES 183

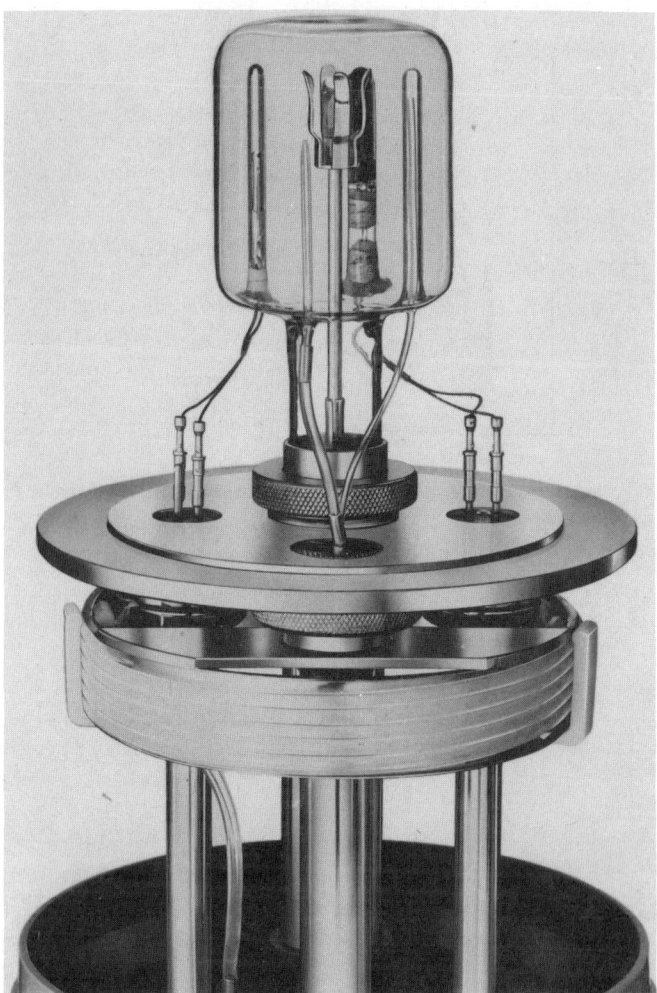

Fig. 9.9. Titration vessel of the LKB titration calorimeter. (Photograph courtesy of LKB Produkter AB, Bromma, Sweden.)

the point of entry into the calorimeter via a thin-walled tube immersed in the bath. The authors state that the electrical calibration has a standard deviation of 0.1% and that results agree with the heat of neutralization to within 0.2%. Precision calorimeters based on the design of Christensen et al.[6] can be obtained from Tronac Inc., Orem, Utah.

Using a differential titration calorimeter, Nakanishi and Fujieda[30] have shown that the difficulties arising during quasiadiabatic (anisothermal) titration (e.g., slow attainment of thermal equilibrium before beginning

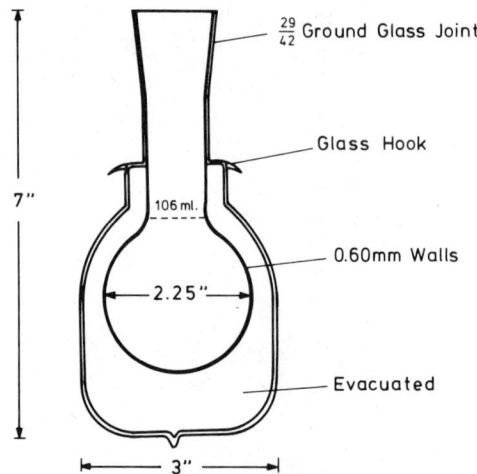

Fig. 9.10. Thermometric titration calorimeter of Christensen, Izatt, and Hansen.[6]

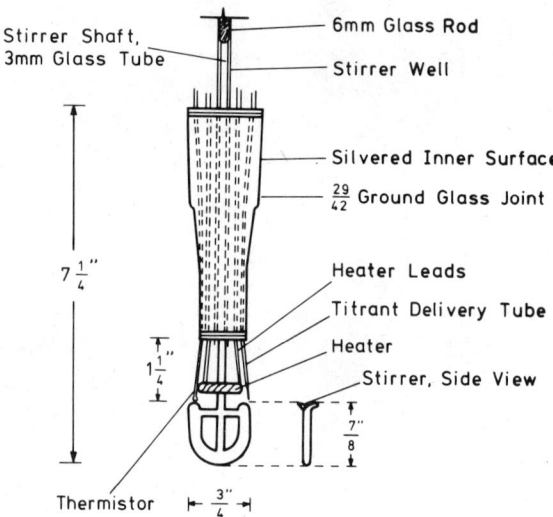

Fig. 9.11. Insert of the calorimeter of Christensen, Izatt, and Hansen.[6]

titration) can be avoided by aiming at isothermal conditions. To this end, both calorimeter vessels of the twin system are immersed in a water bath, thereby obtaining a large constant of heat exchange κ [see (9.1a)] and consequent rapid reaching of thermal equilibrium. Further, the temperature difference Θ between twin reaction and reference vessels remains small when thermal energy is developed in the reaction vessel (chemical reaction or electrical calibration). The output voltage of a Wheatstone bridge containing two matched thermistors is directly proportional to the temperature difference Θ. By means of an analog compensation circuit, this output voltage is converted to a signal directly proportional to the signal that would be expected under fully adiabatic conditions. The capabilities of the method were demonstrated in the titration of hydrochloric acid and of phenol with sodium hydroxide in aqueous solution. The heat of neutralization of -13.46 ± 0.06 kcal mol^{-1} for hydrochloric acid in infinitely dilute solution agrees quite well with the values obtained by others using calorimetric procedures (see Chapter 4). The end point in hydrochloric acid titration could be reproduced to within about 0.2%. These figures show that accurate and precise measurements are possible using the technique proposed by Nakanishi and Fujieda.

The following precision titration calorimeter is described in detail as an example of a system of instrumentation.[31] The construction of this calorimeter (Fig. 9.12) respects the principles discussed in Section 9.1. The cylindrical calorimeter vessel (a) of 200-ml content is made of Pyrex and is based on the development described by Danielsson et al.[27] The vessel is in the lower part (fu) of a bipartite jacket vessel (f) of nickel-plated brass. The upper end of the calorimeter vessel consists of a thin-walled glass tube with a screw plug (c) at its upper extremity. The glass tube is used for fixing the vessel in the lid (b) of the chamber (fu). A thin-walled U-shaped tube for the heater (h) and another tube closed at its lower end, in which the thermistor (t) is placed, are sealed into the calorimeter vessel. The calibration heater (h) (100 Ω) is a manganin wire in the U-shaped tube in Kel-F-oil, which is used for improved heat transfer. The thermistor is embedded in Wood's metal. The leads of heater and thermistor are sealed airtight with epoxy resin. A glass rod (g) at the bottom of the vessel has a Teflon support for the magnetic stirring rod (m), which is driven by a synchronous motor (s) (AEG, type SSLK, 375 rpm) placed under the calorimeter jacket. The screw plug (c) is hermetically sealed with the help of a silicone rubber disk (d), carrying a stainless steel capillary tube (k) (0.32 mm $\emptyset$), through which the titrant is added. At the level of fixing (b') is a Teflon disk (e) in the neck of the calorimeter vessel to reduce convection. This calorimeter vessel was designed for reactions of small reaction power, including those in nonaqueous solution. The calorimeter content is in contact with Pyrex, Teflon, and stainless steel only. The vessel corresponds to the design of Fig. 9.1 and all the calculations made above are valid for the calorimeter described here.

The calorimeter vessel (a) is fixed to the brass lid (b) of the lower chamber (fu) with silver conductive epoxy. The level of fixing (b') defines the boundary of the vessel as described in Section 9.1.1. The lid (b) is used to mount different parts of the calorimeter. It is sealed off from the

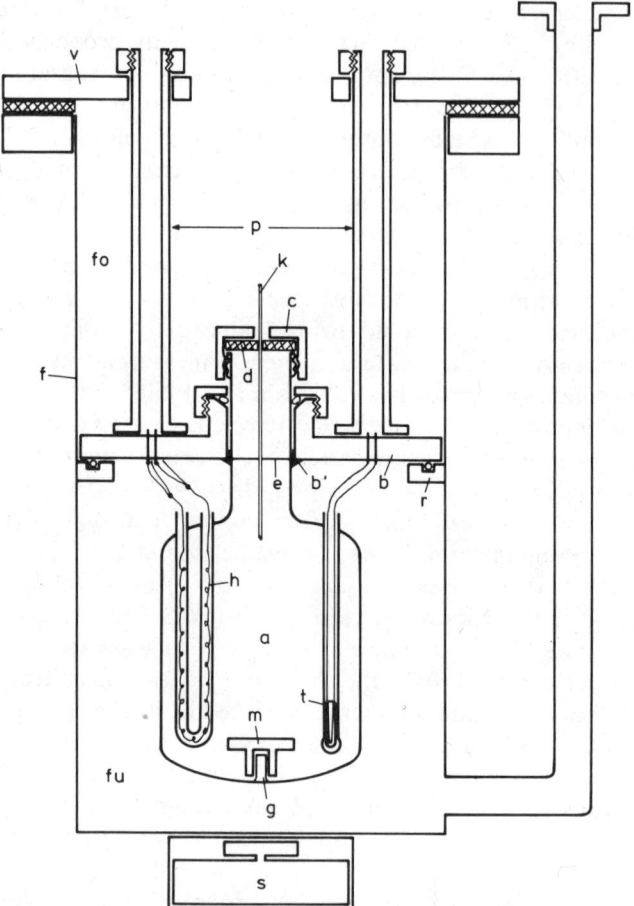

Fig. 9.12. Titration calorimeter.[31]

evacuated lower calorimeter chamber containing the calorimeter vessel with the help of an O-ring placed on a metal flange (r) on the jacket (f). The leads of heater and thermistor pass through the lid vacuum tight; above the lid they are led to the outside through brass tubes (p) fixed to the lid. These tubes function at the same time as supports for a second level of mounting (v) consisting of Plexiglas carrying the burette. The entire jacket vessel is immersed in a thermostated water bath. The upper calorimeter chamber (fo) is thus an air thermostat for the steel capillary for titrant addition and for the burette surrounded by a second thermostat jacket. The air in the upper calorimeter chamber (fo), which is sealed airtight, is dried with silica gel to prevent diffusion of water through the closure (c) of the calorimeter.

The design depicted in Fig. 9.13 aims at a linear and stable system for temperature measurements of high sensitivity. A glass probe thermistor $R(T)$ (Siemens, K17, 10 kΩ) is used as a temperature sensor; its temperature function is linearized according to (9.20).

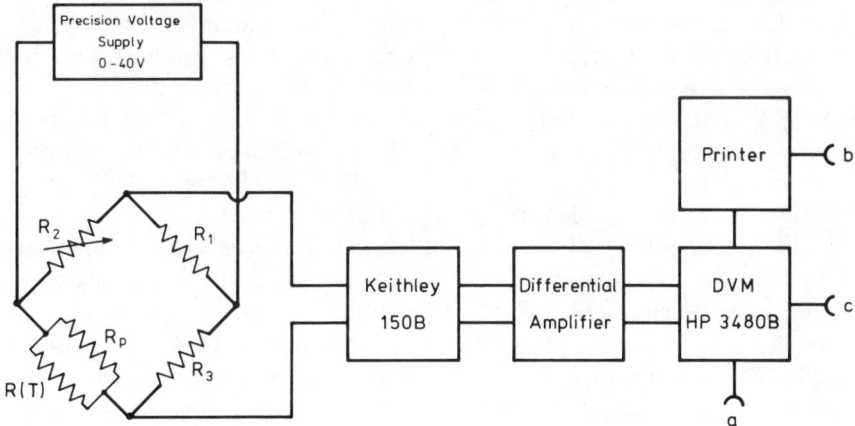

Fig. 9.13. Block diagram of the temperature measuring and recording assembly.[31]

Figure 9.13 shows the construction of the temperature measuring bridge and of the electronic measuring system. The bridge resistances R_1, R_3 (200 kΩ) are of a size guaranteeing constancy of the current to within $3 \cdot 10^{-4}$ in a temperature range of ± 1 K [$\Delta R(T) = \pm 60$ Ω], within which the deviation of the thermistor resistance from linearity is negligibly small; the linearity of the bridge signal is ensured by the constancy of the current flowing through the thermistor. The bridge is supplied by a precision dc voltage source, the output voltage of which can be regulated stepwise between 0 and 40 V. The precision voltage supply has as reference a temperature-compensated Zenerdiode, which guarantees a temperature drift of less than 10^{-5} K^{-1}.

The thermistor is linearized with the help of the resistor R_P (ESI DS 1265, 12 kΩ) as parallel resistance that is easily adapted if the thermistor is changed. The bridge is balanced with the help of a second decade resistor R_2 (ESI DS 1464, 12 kΩ).

The bridge with the resistance values given in Fig. 9.13 is supplied by a voltage of about 40 V; the unbalance voltage is 10 μV/10^{-3} K. The voltage drop at the thermistor under these conditions is 1 V; the power dissipation of the thermistor is, therefore, 0.1 mW. The rise of temperature of the thermistor was experimentally determined to be 0.03 K.

The bridge resistors have temperature coefficients of $5 \cdot 10^{-6}$ K^{-1} and the temperature gradient of the linearized thermistor is about $1.5 \cdot 10^{-2}$ K^{-1}. To measure a temperature rise of 10^{-5} K, the bridge resistors must have a temperature constant to ± 0.03 K, at least during the period of measurement. Elaborate thermostating of the bridge is avoided by using thermal insulation (Styrofoam). Changes of the room temperature (± 0.2 K) then influence the bridge resistors only slowly. The current must not be switched off, however, since the temperature equilibrium within the bridge is established only slowly (power dissipation of the resistors and the voltage source of the bridge). If the temperature of the thermistor sensor changes by 10^{-5} K, the bridge signal changes by only 0.1 µV. A low-noise chopper amplifier (Keithley, model 150B) was therefore used as preamplifier, amplifying the signal in a range of 1 mV by the factor 1000. Since the voltage source is grounded, the preamplifier must not be.

The output voltage of the preamplifier is fed into an isolation amplifier ($R_{in} > 10^{12}$ Ω, voltage gain 10), the output voltage of which is indicated by a digital voltmeter (HP 3480B). This output voltage is registered by a digital printer. The output voltage of the isolation amplifier may also be registered by a recorder. It has already been mentioned in Section 9.1.3 that the burette can be operated manually or automatically.

Automatic thermometric titration is carried out by a digital controlling device, the layout of which is depicted in Fig. 9.14.[31] This controlling device sends an impulse to the digital voltmeter (Fig. 9.13) at equidistant points of time Δt via its output a. The period Δt can be set with a presetting device "measuring period" from 0.1 to 9.9 min (Fig. 9.14). The digital voltmeter shows the unbalance voltage of the temperature measuring bridge after each impulse and, after measuring, gives the order "print" to the storing unit "measuring time and number of titration steps." This order makes the unit print the content of the storing unit "measuring time and number of titration steps" together with the value measured by the digital voltmeter.

The controlling unit fulfills the requirements of a complete thermogram (preperiod, reaction period, and afterperiod) and permits the recording of the preperiod before any titrant is added with the help of its presetting device "preperiod and measuring time." After the predetermined preperiod is over, the titrant is added according to the order given to the unit "stepper motor control." In the presetting device, "motor steps," the number of motor steps, and the amount of titrant to be added have been determined in advance. The presetting device "titration step interval" controls the time interval between the individual steps (the duration of the periods can be set from 0.1 to 9.9 min). The overall number of titration steps is set with the help of the presetting device "titration steps" (1 to 99 steps).

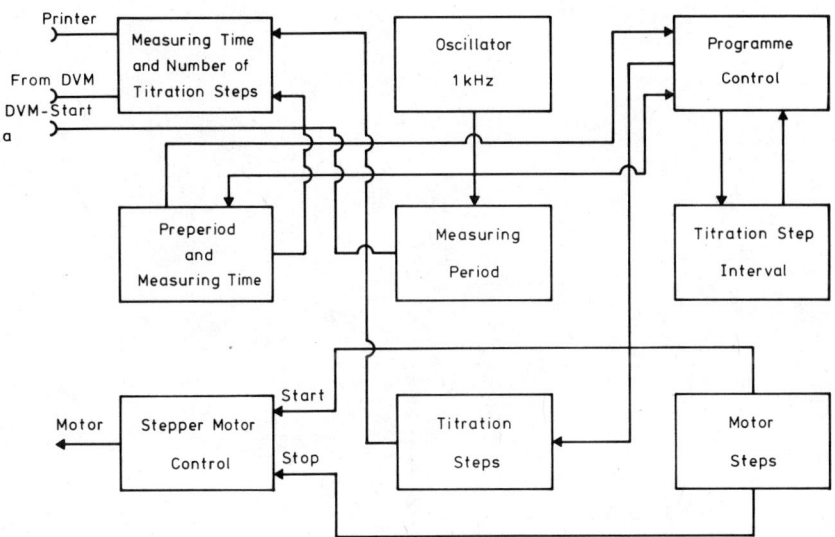

Fig. 9.14. Block diagram of the controlling device of an automatic discontinuous thermometric titrator.[31]

After the last step, only the values of the temperature and the measuring time (afterperiod) are printed until the controlling unit stops. To ensure a compatible setting of the different units, there is a further unit "program control," which indicates incompatible setting.

The temperature Θ of the environment and that of the titrant (T_B) can be controlled accurately only if the temperature of the thermostat bath containing the calorimeter vessel is regulated precisely (see Fig. 9.15). The 40 liters of thermostat liquid are in a stainless steel container insulated with Styrofoam and are circulated by the stirrer (R). The stirrer is in a cylinder (M) containing an electrical heater and a copper tube heat exchanger (W), inserted in the circulation circuit of a commercial thermostat, the temperature of which is below that of the precision thermostat. The water leaving the cylinder (M) at the bottom acquires a uniform temperature in this way. A rotary pump drives the thermostat water through the thermostat jacket of the burette. A heat exchanger W' carries away the power of stirring so that the temperature of the burette is the same as that of the thermostat liquid to within $\pm 10^{-3}$ K. The temperature of the commercial thermostat is regulated so that the necessary electrical heating power is about 20 to 40 W. The heater is governed by a PID controller, which is regulated by a thermistor as temperature sensor in a Wheatstone bridge. PID controllers have good dynamic properties and the advantage over P controllers that their mean deviation is zero. The short-term temperature

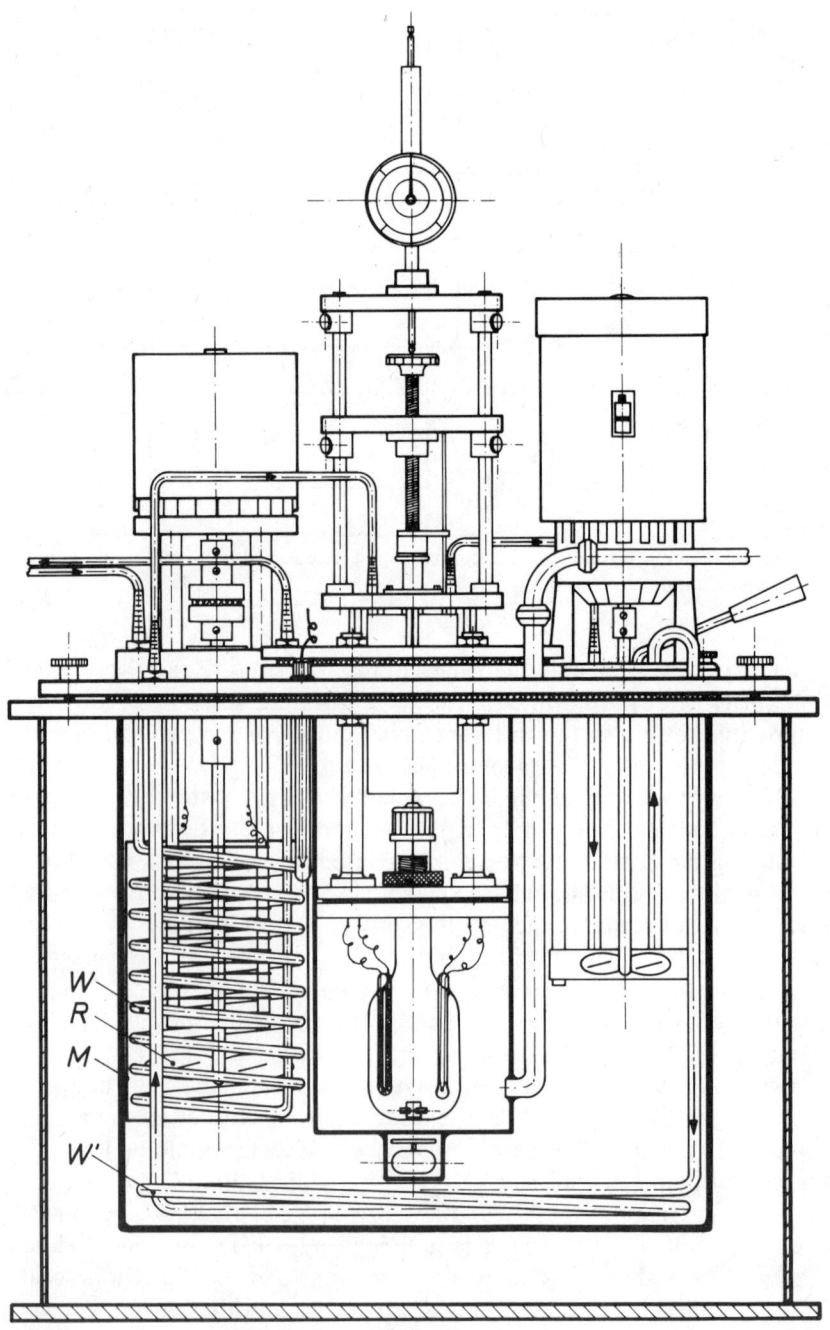

Fig. 9.15. Thermometric titration assembly.[31]

fluctuations of the thermostat liquid are of the order of $\pm 3 \cdot 10^{-4}$ K; the long-term drift is about $5 \cdot 10^{-4}$ K per 24 hr.

The heating circuit corresponds more or less to that depicted in Fig. 9.8. The electrical heater in the calorimeter vessel (manganin wire of 100 Ω) is supplied by an electronic voltage source (Gossen, type 24 KR 0.8). A stopwatch (Zivy, type 309e, accuracy 0.01 sec) is switched on at the same time as the current. The voltage drop E_H across the heater is measured via potential leads, using a high-impedance digital voltmeter (HP 3480B with buffer amplifier 3481A). The resistance R_H of the heater was previously measured with the help of a Kelvin bridge (ESI, type PVB 300A) and the heating power calculated according to

$$W = \frac{E_H^2}{R_H} \qquad (9.24)$$

The heating current can also be determined by measuring the voltage drop across a standard resistor R_S (1Ω) inserted into the heating circuit.

9.2.2 TITRATION CALORIMETERS FOR PREDOMINANTLY ANALYTICAL WORK

Analytical work aims at a determination—as fast and as accurate as possible—of the end point of a chemical reaction. A determination of the heat of reaction is of incidental interest only. The TET method (see Section 3.1) concentrates on the change of temperature during the titration; the DIE method (see Section 3.2) is concerned with the instantaneous "jump" of the temperature; and the continuous flow method determines the stationary temperature difference established after a certain time (see Section 3.3). The methodological basis of the different processes has been discussed in Chapter 3.

The TET Method. In the TET method the end point is indicated by the change of the temperature rise dT/dt. As a rule it is determined by graphical extrapolation. The first commercially made titrator providing automatic recording of a thermogram was the Titra-Thermo-Mat of the American Instrument Company, Silver Spring, Maryland. It is based on Jordan's fundamental design.[21] The end point is determined manually on the recorded thermogram by graphical extrapolation. A method with the goal of automatic end-point indication can start from the assumption that the rise in temperature of the titration period is large compared to those of pre- and afterperiods.

The first derivative of the function $T = f(t) = f(v)$ (thermogram) results in a rectangularlike shape if the titration reaction is fast and complete.

During the preperiod, $dT/dt \approx 0$; during the reaction period, $dT/dt \approx \text{const} \gg 0$; and after the end point, again $dT/dt \approx 0$. The second derivative of the function $T = f(t)$, therefore, consists of two peaks at the beginning and the end of the reaction; from their distance apart the amount of titrant solution used can be determined.[32]

An automatic end-point detector described by Priestley[33] uses the first derivative for the automatic end-point indication. The block diagram of Fig. 9.16 illustrates the working of this titrator. The output voltage of the thermistor bridge is first smoothed in order to suppress the pickup of the line voltage. The signal is then differentiated with the help of an R-C circuit. The amplitude of the differentiated signal, which is directly proportional to the temperature rise dT/dt, is amplified (by the factor 1000) and fed into a discriminator circuit. This is set so that, at a predefined voltage characterizing the beginning and the end of the reaction, it switches on and off, respectively. The motor-driven burette for titrant addition is started manually. As soon as the temperature begins to rise, the end-point detector switches on an electric counter generating an impulse every 0.01 sec. When the end point is reached, the counter and burette are switched off.

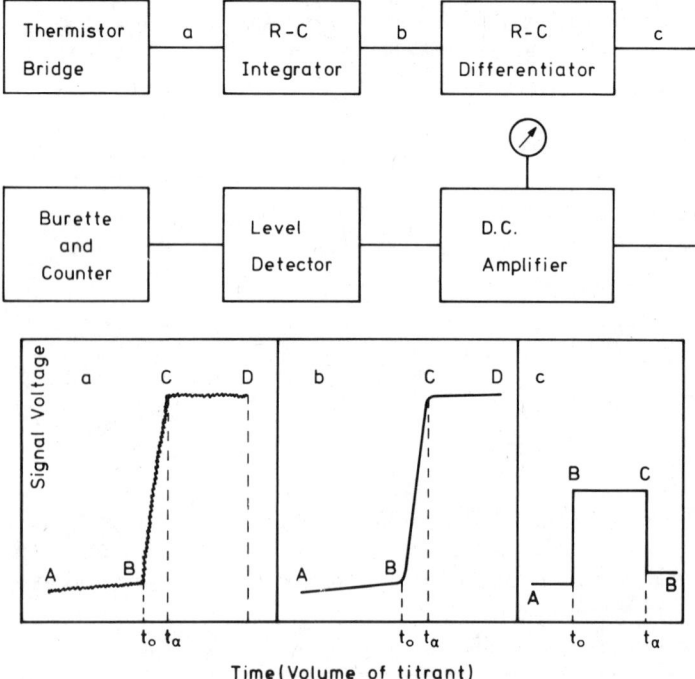

Fig. 9.16. Block diagram of Priestley's automatic end-point titrator.[33]

The number indicated by the counter is directly proportional to the amount of titrant used. This titrator can be used if the temperature rise is sufficient (at least 0.01 K sec^{-1}). The large power of reaction necessary for this is obtained by the fast addition of the titrant (7 ml min^{-1}). An effective vibration stirrer ensures the fast mixing of the reactants. The accuracy obtained with such a titrator ranges from 0.3 to 3%.

De Leo and Stern[34] described a method for automatic indication of the end point in which the output signal of the bridge is first amplified by a servomechanism. The amplified signal is then filtered and differentiated. The differentiated signal of a thermometric titration resembles that of a potentiometric titration and can, therefore, be fed directly into an automatic potentiometric titrator (Sargent–Malmstadt).

The DIE Method. The DIE method is particularly useful for automatic routine analysis. The temperature rise after the addition of a sample solution is directly proportional to the amount of substance undergoing reaction. It is possible, therefore, by a series of comparative measurements, to recalibrate as a concentration scale the unbalance voltage of the bridge or the deflection of the recorder, or, if the bridge is balanced, the change of the thermistor resistance. Direct indication of percentages is also feasible.[35] A prerequisite for these methods is that the reactions take place under standard conditions. The method as originally used[36] minimizes the errors arising from the heat of mixing, due to different temperatures of the solutions of the reactants, by adding a small but concentrated amount of the reacting solution to the sample.

On the other hand Sajó and Sipos[35] use an appropriately designed pipette, which dips the reacting solution into the sample in the calorimeter vessel (Fig. 9.17). After temperature equilibrium has been reached, the reacting solution is expelled into the sample solution. The temperature rise of the reaction is corrected by an amount due to the heat exchange of the calorimeter with its environment. This method avoids errors due to different temperatures of the reactant solutions. Sajó and Sipos used this thermometric method for the analysis of slag,[35] fire clay, kaolin, and clay,[37] and for the determination of impurities in metals.[38] It has opened up a wide field of industrial application for this method.

The analysis apparatus Directthermon, constructed on the basis of Sajó's method by the Hungarian Optical Works, Budapest, Hungary, makes this method of analysis available for routine laboratory work.

Doering[39] has described the use of this apparatus in glass analysis. The Wheatstone bridge of the Directthermon makes it possible to carry out thermometric analyses by calibrating the scale of the instrument with the help of test reactions. Direct indication of the concentration value is also possible.

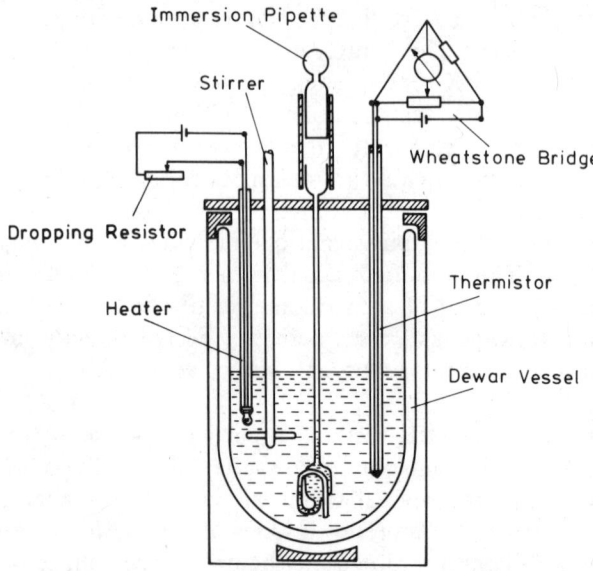

Fig. 9.17. Sketch of titration equipment of Sajó and Sipos.[35]

Guillot[40] has described an apparatus for automatic thermometric titration by the DIE method. Figure 9.18 shows the design of this apparatus. Its essential parts are the Plexiglas calorimeter vessel insulated by Styrofoam, arrangements for the automatic addition of the reactants and for emptying the calorimeter vessel, devices for control of the measuring process and for registering the temperature. The pneumatic pipette device is adjusted so that the titrant is present in excess. The subsequent procedure is automatic: The reacting solution is fed from the supply chamber into the calorimeter vessel. The sample is drawn in; after thermal equilibrium has been established and the bridge balanced, the sample is injected, the temperature rise registered and, finally, the calorimeter vessel emptied. This whole measurement procedure is completed within 3 min. The author reports an accuracy of 1% for a number of neutralization, redox, and complex formation reactions. As the titration process is fully automatic, this titrator can be used for process control.

The Continuous-Flow Method. Genuine on-line control is possible only with the continuous-flow method. The method was developed by Priestley, Sebborn, and Selman[41] and is described in Section 3.3.1. The difference between the temperature of the end-product solution and the mean temperature of the two reacting solutions is determined with the help of a linearized thermistor bridge.[23] Štráfelda and Kroftová,[42] in contrast, allow

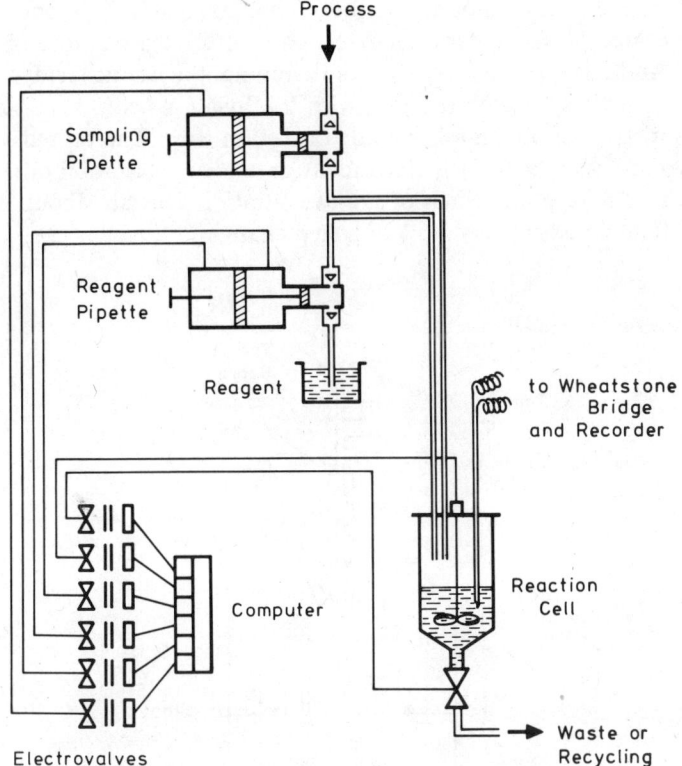

Fig. 9.18. Guillot's automatic DIE titrator.[40]

the two reactants to flow through a heat-exchanger first to bring both to the same temperature. The reactants then flow through closely adjacent tubes, which are in contact with the cold junctions of a thermopile. The two solutions are then mixed and flow through a tube along the warm junctions out of the calorimeter. The properties of the analyzer were studied with the aid of the heat of reaction of sodium hydroxide with hydrochloric acid. The output of the analyzer is directly proportional to the concentration of the acid to be determined.

Continuous-flow analyzers for on-line work are also described by Snyder.[43] The determination of the NaOH content of a nonaqueous reaction mixture, using a solution of 16% CH_3COOH in isopropanol, is carried out in a Teflon titration chamber fitted with a stirrer (Fig. 9.19). The reactants flow into the chamber through a single opening. The reference thermistor above the point of entry is situated so that it measures the temperature of the sample solution; the measuring thermistor is fitted in the top of the reaction chamber. The titration cell is enclosed in a thermostated jacket. Heat exchangers control the temperature of the solutions. Another titration cell for determining the acetic acid content of organic solvents through reaction with triethylamine consists essentially of a T-shaped reaction chamber made of Teflon tube. The output voltage of the two analyzers is linearly dependent on the concentration of the sample solution. The response times of the two titration cells are about 15 min, or 25 to 30 min, respectively (95% of a step change).

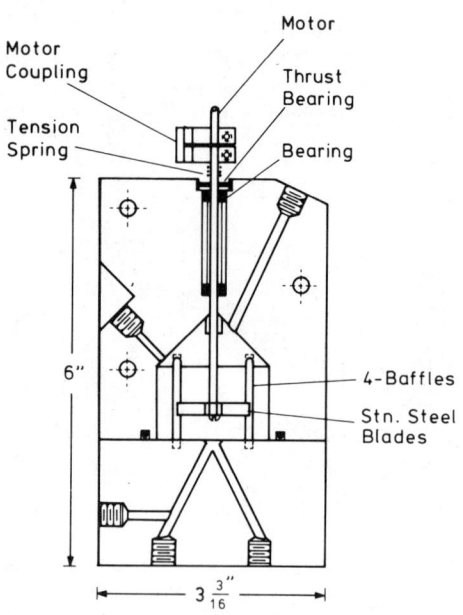

Fig. 9.19. Titration cell of Snyder's onstream hydroxide analyzer.[43]

A continuous-flow analyzer designed for the measurement of large temperature changes is described by Taubinger.[44] Sample and reagent solutions flow through heat-exchanger coils, which are placed in a water-filled thermostat, before entering the mixing chamber. The temperature rise at the exit of the mixer is measured with the help of a thermistor bridge circuit and recorded by a strip-chart recorder.

9.2.3 MICROCALORIMETERS

Highly elaborate titration calorimeters[5,6] have a sensitivity of $2 \cdot 10^{-5}$ K, a heat capacity C of about 100 cal K^{-1}, and an accuracy of $2 \cdot 10^{-3}$ cal in determining the heats of reactions. A reaction enthalpy ΔH of 10 kcal mol^{-1} corresponds to $2 \cdot 10^{-7}$ mol of substance. As it is not possible to measure temperature changes of less than about 10^{-5} K, smaller amounts of substances can be determined with the TET and the DIE method only if a smaller volume of the reaction solution is used. Jordan, Henry, and Wasilewski[45] describe a microcell containing 2000 µl, used in combination with the industrial Titra-Thermo-Mat apparatus. The sensitivity of this experimental setup is, however, not better than that of a precision macrocalorimeter because the temperature resolution is limited to 10^{-3} K.

Special precision microcalorimeters, working either by the flow calorimetric method[46,47] or by the batch calorimetric method,[48] are sensitive to 10^{-7} cal sec^{-1} (flow calorimeters) or 10^{-6} cal (batch calorimeters). In the batch method, the solutions of the two reactants are in two separate chambers of the reaction vessel. The solutions are mixed by revolving the vessel. The measuring principle is that of the DIE method.

Microcalorimeters belong to the heat-conduction type. The heat exchange between reaction vessel and the surrounding heat sink takes place via a thermopile until $\Delta T = T - \Theta$ equals zero. The junctions of the thermopile are suitably distributed over the surface of the reaction vessel and the surrounding heat sink so that it is possible to measure the integral heat flow W (cal sec^{-1}). This is directly proportional to the mean temperature difference ΔT

$$W = K \Delta T$$

in which the heat transfer coefficient K (cal sec^{-1} K^{-1}) is a constant of the apparatus. The emf E of the thermopile is directly proportional to the temperature difference ΔT. The amount of heat Q exchanged during the entire period of measuring t(sec) (up to the establishment of temperature equilibrium) equals

$$Q = K_1 \int_0^t E(t) \, dt$$

in which K_1 is again a constant of the apparatus to be determined by electrical or chemical calibration. This principle of measurement, first applied by Tian and Calvet,[7] has the advantage over temperature measurement that the temperature in the calorimeter vessel does not have to be homogeneously distributed. It is, for example, unnecessary in the batch method to stir the reaction solution; thus the heat of stirring that can upset the determination of small heats of reaction is avoided. Precision microcalorimeters are, as a rule, designed as twin calorimeters. It thus becomes possible to diminish effects of the environment, which applies equally to both calorimeter systems.

The main field of application of precision microcalorimeters is in biochemical investigations. For instance, Monk and Wadsö[49] determined with the aid of a flow microcalorimeter the concentration of glucose and the activities of glucose oxidase, cholinesterase, alkaline phosphatase, lactic acid dehydrogenase, and APTase in a tissue homogenate. Precision microcalorimeters are produced by LKB Produkter AB, Sweden (flow microcalorimeter LKB 10700-1, batch microcalorimeter LKB 10700-2).

REFERENCES

1. M. Jakob and G. A. Hawkins, *Elements of Heat Transfer*, 3rd ed., Wiley, New York, 1967.
2. F. Kohlrausch, *Praktische Physik*, Vol. 3, B. G. Teubner, Stuttgart, 1968.
3. I. Wadsö, *Sci. Tools*, Vol. 13, No. 3, 33 (1966).
4. R. V. Churchill, *Operational Mathematics*, 3rd ed., McGraw-Hill, New York, 1972.
5. S. Sunner and I. Wadsö, *Acta Chem. Scand.*, **13**, 97 (1959).
6. J. J. Christensen, R. M. Izatt, and L. D. Hansen, *Rev. Sci. Instrum.*, **36**, 779 (1965).
7. E. Calvet and H. Prat, *Recent Progress in Microcalorimetry*, edited and transl. from French by H. A. Skinner, Pergamon Press, Elmsford, N. Y., 1963.
8. W. Walisch and F. Becker, *Z. Phys. Chem.*, (New Series), **46**, 268 (1965).
9. W. Walisch and F. Becker, *Z. Phys. Chem.* (New Series), **46**, 279 (1965).
10. J. J. Christensen, H. D. Johnston, and R. M. Izatt, *Rev. Sci. Instrum.*, **39**, 1356 (1968).
11. W. V. Johnston and G. W. Lindberg, *Rev. Sci. Instrum.*, **39**, 1925 (1968).
12. G. Bosson, F. Gutmann, and L. M. Simmons, *J. Appl. Phys.*, **21**, 1267 (1950).
13. K. Schlyter, *Trans. R. Inst. Technol. (Stockh.)*, **132**, (1959).
14. *Thermistor Technical Data*, TD-1, Fenwal Electronics, Inc.
15. J. Jordan, in I. M. Kolthoff and P. J. Elving, *Treatise on Analytical Chemistry*, Vol. 8, Interscience, New York, 1968, Part I, p. 5175.
16. H. W. Linde, L. B. Rogers, and D. N. Hume, *Anal. Chem.*, **25**, 404 (1953).
17. P. J. Reilly and L. G. Hepler, *J. Chem. Educ.*, **49**, 514 (1972).
18. B. C. Tyson, Jr., W. H. McCurdy, Jr., and C. E. Bricker, *Anal. Chem.*, **33**, 1640 (1961).
19. P. Nordon and N. W. Bainbridge, *J. Sci. Instrum.*, **39**, 399 (1962).
20. H. Hahn, *Thermistoren*, R. v. Decker's Verlag, G. Schenck, Hamburg, 1965.
21. J. Jordan and T. G. Alleman, *Anal. Chem.*, **29**, 9 (1957).
22. W. L. Everson, *Anal. Chem.*, **39**, 1894 (1967).
23. P. T. Priestley, *J. Sci. Instrum.*, **42**, 35 (1965).
24. P. Gerding, I. Leden, and S. Sunner, *Acta Chem. Scand.*, **17**, 2190 (1963).
25. P. Dutoit and E. Grobet, *J. Chim. Phys.*, **19**, 324 (1922).
26. J. Barthel and N. G. Schmahl, *Z. Anal. Chem.*, **207**, 81 (1965).
27. I. Danielsson, B. Nelander, S. Sunner, and I. Wadsö, *Acta Chem. Scand.*, **18**, 995 (1964).
28. J. J. Lingane, *Anal. Chem.*, **20**, 285 (1948).

29. M. J. Stern, R. Withnell, and R. J. Raffa, *Anal. Chem.*, **38**, 1275 (1966).
30. M. Nakanishi and S. Fujieda, *Anal. Chem.*, **44**, 574 (1972).
31. R. Wachter, J. Barthel, and K. Wachter-Lenz, in preparation.
32. S. T. Zenchelsky and P. R. Segatto, *Anal. Chem.*, **29**, 1856 (1957).
33. P. T. Priestley, *Analyst*, **88**, 194 (1963).
34. A. B. De Leo and M. J. Stern, *J. Pharm. Sci.*, **54**, 911 (1965).
35. I. Sajó and B. Sipos, *Z. Anal. Chem.*, **222**, 23 (1966).
36. J. C. Wasilewski, P. T. -S. Pei, and J. Jordan, *Anal. Chem.*, **36**, 2131 (1964).
37. I. Sajó and B. Sipos, *Tonind.-Ztg.*, **92**, 88 (1968).
38. I. Sajó and B. Sipos, *Mikrochim. Acta*, **1967**, 248.
39. K. Doering, *Silikattechnik*, **21**, 169 (1970).
40. P. Guillot, *Anal. Chim. Acta*, **50**, 499 (1970).
41. P. T. Priestley, W. S. Sebborn, and R. F. W. Selman, *Analyst*, **90**, 589 (1965).
42. F. Štráfelda and J. Kroftová, *Collect. Czech. Chem. Commun.*, **33**, 3694 (1968).
43. K. L. Snyder, *Chem. Eng. Prog.*, **64**, 75 (1968).
44. R. P. Taubinger, *Analyst*, **94**, 634 (1969).
45. J. Jordan, R. A. Henry, and J. C. Wasilewski, *Microchem. J.*, **10**, 260 (1966).
46. P. R. Stoesser and S. J. Gill, *Rev. Sci. Instrum.*, **38**, 422 (1967).
47. P. Monk and I. Wadsö, *Acta Chem. Scand.*, **22**, 1842 (1968).
48. I. Wadsö, *Acta Chem. Scand.*, **22**, 927 (1968).
49. P. Monk and I. Wadsö, *Acta Chem. Scand.*, **23**, 29 (1969).

INDEX

Acetamide, 103
Acetate, determination of, in glacial acetic acid, 103
  metal complexes of, 148
Acetic acid, as solvent, 21, 98, 99, 102, 103, 105, 106, 109, 132
  determination of, 72, 84, 85, 104
    in acetic anhydride, 99
    in acetylating mixtures, 108
    in mixtures of acids, 87, 93
  thermodynamic data for neutralization of, 84
  see also Acetic anhydride; Water
Acetic anhydride, acid catalysed hydrolysis of, 99, 105
  as 'enthalpimetric indicator' solvent, 106
  content of, in acetic acid, 99
  see also Acetic acid
Acetone, as "enthalpimetric indicator" solvent, 104-105
  as solvent, 132
  in binary mixture, enthalpy of mixing, 65
Acetonitrile, as solvent, 103
Acetylacetone, 105
Acetylating mixtures, 108-109
Acetyl value, 2, 109
Acrylonitrile, 72
  as "enthalpimetric indicator" solvent, 106-107
Actinides, sulfate complexes of, 149; see also Thorium; Uranium
Activity coefficient, 9, 15-17
Activity function, 8
Addition of titrant solution, 61-63, 176-177
  discontinuous, automatic device for, 188-189
Adipic acid, 91
Alcohols, determination of, in hydrocarbons, 99
Alkaline phosphatase, 73
Alkaloids, 107
  salts of, 107
Alkylphenols, see Hydroxyl value
Alloys, analysis of, 133, 149
Alumina-silica gels, as Lewis acids, 99
Aluminum, basic complex salts of, 120, 121
  determination of, 119, 120, 121, 123, 149, 151
  determination of its natural impurities, 75
  fluoride complexes of, 119-120, 149
  organic acid chelates of, 150, 151
Aluminum alkyls, determination of, 101
Aluminum tribromide, as Lewis acid, 99
Aluminum trichloride, as Lewis acid, 99

Amides, 87, 107
Amines, as titrants for organoaluminum compounds, 101
  determination of, 85, 104, 105, 107, 108
  thermodynamic data for neutralization of, 104
Amino acids, determination of, 84, 87, 95
Aminocarboxylates, 150
2-Aminoethanol, mercury(II)complex of, 147
α-Aminoisobutyric acid, 148
Aminopyrine, 106
Ammine, of copper, 139
Ammonia, determination of, 62, 85, 88, 120
  in presence of amines, 88
Ammonium borate, 85
Ammonium chloride, determination of, 88
  in mixture with amine hydrochlorides, 88
Ammonium citrate, 77
Ammonium ferrous sulfate, 128-130
Ammonium sodium hydrogen orthophosphate, 123
Anilides, 108
Aniline, 88, 108
  derivatives of, 107
Antimony, fluoride complexes of, 149
Antipyrine, 106
Arsenic(III), in "enthalpimetric indicator" reactions, 124
Arsenic acid, determination of, 93
  in mixtures of acids, 93
Arsenious acid, as titrant, 133
  determination of, 93, 129, 133
Ascorbic acid, 133
ATPase, 73
Autoprotolysis, of amphiprotic solvents, 20
  constant, 15, 18, 20
  of water, 15
Azelaic acid, 91
Azide complexes of Cadmium, 148

Barium, determination of, 115-117, 122, 151
  organic acid chelate of, 151
Barium acetate, as titrant, 109
  titration with perchloric acid, 103
Bayer process solution, 94
Benzene, as solvent, 7, 99, 100, 101, 104
  determination of, 70, 108
Benzoic acid, determination of, 104, 107
  derivatives of, 107
  thermodynamic data, 104
p-Benzoquinone-amidinohydrazone-thiosemicarbazone, 106

201

Benzylthiuronium chloride, 124
Beryllium, basic salts of, 136
  determination of, 151
  fluoride complexes of, 136, 137
  organic acid chelate of, 151
  pyrophosphate complexes of, 136
Binary mixtures, heat of mixing of, 64-66
  principles of analyzing of, 53
2,2'-Bipyridyl, as titrant for organometallic compounds, 101
Bismuth, organic acid chelate of, 151
Bisulfite, see Hydrogen sulfite
Blast furnace slags, see Slags
Boric acid, destruction of fluoride complexes by, 149
  determination of, 3, 21, 39, 40, 67, 72, 84
  in mixtures of acids, 86, 87
  thermodynamic data for neutralization of, 83, 84
Boron trifluoride, as titrant, 107
Brass, analysis of, 149
Bromide, as titrant, 115
  mercury, complexes of, 147
  determination of, 118, 124
Bromine water, 54
p-Bromoaniline, 104
Burette, 57, 176-177
  thermostating of, 176-177
Butanol, as titrant for butyl lithium, 101
n-Butylamine, 104
Butyl disulfide, 101
tert-Butyl disulfide, 101
Butyl lithium, 101
n-Butyric acid, in acid mixtures, 87
  determination of, 84
  thermodynamic data for neutralization of, 84

Cadmium, acetate complexes of, 148
  azide complexes of, 148
  basic sulfates of, 136
  cyanide complexes of, 148
  determination of, 120, 151
  halide complexes of, 136, 137, 147
  hexacyanoferrate compounds of, 119, 136
  nitrite complexes of, 148
  organic acid chelates of, 150, 151
  sulfate system, association in, 141
  thiocyanate complexes of, 148
  thiosulfate complex of, 139
Cadmium acetate, titration with perchloric acid, 103
Cadmium chloride, complex formation in a KCl-CdCl$_2$ melt, 109
Cadmium perchlorate, hydrolysis equilibria of, 148
Calcium, determination of, 3, 119, 122, 151
  alongside magnesium, 3, 123, 152
Calibration of titration calorimeters, 177-181
  electrical circuit for, 179
  electrical heaters for, 179-180
  by electrical heating, 179-181
  standard reactions for, 77-80, 81
  by standard reactions, 82, 181
Calorimeter, adiabatic, 167
  constant-temperature, 167
  constant-temperature environment, 57-58, 166
  Dewar vessel, commerical, 57-58, 166, 181
  Dewar vessel, of proper design, 182
  differential, 183-185
  vacuum-jacket, 182
Calorimetric equipment, for accurate enthalpy change measurement, 181-191
  for predominantly analytical work, 56-58, 191-194
  for process control, 194-196
  see also Direct injection enthalpimetry; Continuous-flow enthalpimetry; Thermometric enthalpy titration
Carbazole, 108
Carbonate, as titrant, 123
  determination of, 94
Carbon dioxide, 7
Carbon tetrachloride, as solvent, 99, 104
Carboxylic acids, determination of, 84, 87, 89-91, 105, 150-152
  metal complexes of, 150-152
Carboxylic acid anhydrides, determination of, in presence of the parent acids, 99
Catalytic thermometric titration, 5, 75, 104-107, 124, 152
Cellulose, acetylation of, 109
Cements, analysis of, 69, 123, 133
Cerium, as titrant, 62, 129, 131
  determination of, 62, 128, 129, 151
  hexacyanoferrate compound of, 137
  organic acid chelate of, 151
Chemical equilibrium, 12-13
Chemical potential, 8, 16
Chloride, as titrant, 62, 115
  determination of, 62, 118, 124
  metal complexes of, 148
Chlorine, 133
m-Chloroaniline, 104
Chlorobenzene, as solvent, 104
Chloroform, as solvent, 99
  in binary mixture, enthalpy of mixing, 65
Chloropheniramine maleate, 95
Chloropromazine hydrochloride, 95
Cholinesterase, 73
Chromium, as titrant, 128
  determination of, 95, 128, 130, 132, 133
  organic acid chelate of, 151
  polyanion formation, 95
Chromium plating solutions, see Plating solutions
Chromotropic acid, mixed ligand chelates, 150
Cinchonine, 106
Citric acid, as chelating agent, 150
  determination of, 89-90, 150
  thermodynamic data for neutralization of, 90-91
Clay, analysis of, 69, 119, 133, 149

INDEX 203

Clinker, analysis of, 133
Coal, water content in, 2, 70
Cobal(II), cyanide complexes of, 119, 148
  determination of, 74, 119, 151, 152
  hexacyanoferrates of, 119, 136
  organic acid chelates of, 151, 152
  pyrophosphate complexes of, 137
Codeine phosphate, 95
Computer program, 146, 147, 149
Constant of heat exchange, components of, 159-163
  definition of, 33
  dependence on time of, 164
  determination of, 60, 163
Continuous-flow analyzers, 194-196
Continuous-flow enthalpimetry, as on-line process control, 72
  basis of, 70-71
  description of, 70
  equipment for, 70, 72, 194-196
  flow techniques on premixed reaction systems, 72-73
  in microcalorimetry, 72, 197, 198
  stopped flow technique, 72
Cooling constant, see Constant of heat exchange
Copper(I), determination of, 128, 130
  thiosulfate complex of, 137
Copper(II), acetate complexes of, 148
  amine complexes of, 147
  ammine complex of, 139
  amino-acid complexes of, 148
  basic salts of, 136
  cyanide complexes of, 149
  determination of, 123, 133, 149, 151, 152
  fluoride complex of, 149
  hexacyanoferrate compounds of, 136
  organic acid chelates of, 150, 151, 152
  sulfate system, association in, 141
Copper perchlorate, hydrolysis equilibria of, 141
Cresols, bromo derivatives of, 104
  determination of, 84, 85
Cyanide, determination of, 118-119, 124
  metal complexes of, 119, 148, 149
Cyanuric acids, 107
Cyclohexane, as solvent, 70, 101, 108
Cyclohexylamine, 88
Cysteine, 141

DCTA, 107, 152
  mixed ligand chelates, 150
1,2-Diaminocyclohexane-N,N,N',N'-tetra-acetate, see DCTA
Diammonium hydrogen orthophosphate, as titrant for magnesium, 123
Diazonium salts, as titrants for phenols, naphtols, cresols, 85
Dibutyl sulfide, as titrant for iodine, 101
Dibutylthioethane, as Lewis base, 100
DIE, see Direct injection enthalpimetry
Diethoxyethane, as Lewis base, 100
Diethylamine, 88

Diethylzinc, 101
N,N-Dimethylacetamide, iodine complex of, 101
Dimethylformamide, reaction of, with phenol, 13, 141
Dimethyl sulfoxide, as solvent, 107, 152
1,3-Dinitrobenzene, 132
2,4-Dinitrobenzoic acid, 132
Dioxane, as Lewis base, 7, 99
1,3-Diphenylguanidine, as titrant for acids, 104
Diphosphoric acid, determination of, 93
  in mixture of phosphoric acids, 93
Direct injection enthalpimetry, definition of, 1
  description of, 66-67
  equipment for, 193-194
  in microanalysis, 69
  in multicomponent systems, 54
  theoretical basis of, 68-69
  with direct percentage-reading equipment, 69, 119, 193
DIRECTTHERMON, 193
Dissociation constant, of acids and bases, 82-93
  definition of, 15, 18, 41
Disulfides, iodine complexes of, 101
1,2-Dithiane, iodine complex of, 101
Dolomite, analysis of, 69, 123, 133

EDTA, as titrant, 3, 67, 69, 107, 150-152
  metal chelates of, 151
  mixed-ligand chelates, 150
  properties of, 150
  thermodynamic data for chelate formation, 151
Electrical equipment, for calibration, 179-181
  for temperature measurement, 171-176
Electrical resistance thermometers, 169-170
  temperature coefficient of, 169
Emissivities, 161
Emissivity factor, 160, 161
Endpoint, accuracy, 53, 61-63, 122, 151
  amplification through catalytic thermometric titration, 75, 104-107, 124, 152
  indication, spurious, 39, 42, 43, 93
  influence of equilibrium on, 40-42
  intensification through high heat of dilution, 75, 107-108
  principles for calculation of, 38-43
End-point detector, automatic, 192-193
Enthalpimetric indicator, 105
Enthalpy of dilution, of aqueous solutions, 77-80
  calculation from thermogram, 37-53
    by section method, 43-46
    by tangent method, 46-47
  definition of, 32
Enthalpy of dissociation, weak acids, 83-87, 89-92, 116-117
Enthalpy of hydration, 114, 115
Enthalpy of mixing, binary systems, 65

ternary systems, 66
Enthalpy of neutralization, as standard for calibration, 72, 82, 181
  calorimetric determination of, 77-80, 81-82
  concentration dependence of, 77-80
  definition of, 77
  electrochemical determination of, 80
  influence of ionic strength on, 82
  temperature dependence of, 80
Enthalpy of precipitation, 113-118, 119, 122, 123
  concentration dependence of, 114-115
Enthalpy of reaction, approximative calculation of, 38-40
  calculation from thermogram of, 35-53, 66-69
    by initial slope method, 51-52
    by section method, 43-46
    by tangent method, 46-47
  definition of, 7
  pressure dependence of, 11
  temperature dependence of, 11
  see also Thermogram
Enthalpy of redox reactions, 127-130
  concentration dependence of, 131
Enthalpy titration, 1
Entropy of reaction, calculation of, 47-49, 140-149
  definition of, 7
  temperature dependence of, 11
Entropy titration, 50, 100, 139, 140-149
Enzyme activities, determination of, 73
Ephedrine, 141
Equilibrium constant, calculation from thermogram, 47-49, 139-140, 140-149
  definition of, 4, 12, 15
  influence of, on end-point indication, 13, 40-42, 53
Equilibrium time of calorimeters, determination of, 165
  estimation of, 163-166
Ethanol, as solvent, 132
Ethanolamine, determination of, 88
  in mixtures of ammonia and amines, 88
Ethers, as titrants, 99, 100, 101
Ethyl acetate, as Lewis base, 99
Ethyl disulfide, 101
Ethylenediamine, determination of, 88
  reaction of, with mercury chloride, 148
Ethylenediaminetetraacetic acid, see EDTA
Ethylene dichloride, as solvent, 101

Fabric-bleaching solutions, 133
Fats, acetyl value of, 2, 109
  iodine value of, 2, 108
Ferrous ethylenediammonium sulfate, 131
Ferrous sulfate, 130
Fluoride, complexes of, 119-120, 136, 137, 148, 149
  as titrant, 119-120, 149
  destruction by boric acid, 149
  determination of, 119, 149
Fluorocomplex ions, see Fluoride, complexes of
Flow microcalorimetry, 72-73, 197, 198
Formaldehyde, 133
Free-acid content, determination in presence of hydrolyzable cations, 87, 88

Fumaric acid, 91

Gallium, organic acid chelates of, 150, 151
Gibbs energy of reaction, definition of, 12
  determination from thermogram of, 47-49, 140-149
  pressure dependence of, 12
  temperature dependence of, 12
  see also Equilibrium constant
Gibbs energy titration, 2, 53
Glacial acetic acid, as solvent, see Acetic acid
Glucose, 73
Glucose oxidase, 73
Glutaric acid, 91
Glycine, metal complexes of, 148
  protonation of, 148
  reaction with mercury chloride, 148
Grignard reagent activity, 102

Heat capacity, molar, 11
  of calorimeter, 6, 30, 38, 58-60, 177-181
  specific, 30, 58-60
Heat of, see also Enthalpy of
Heat of exchange, see Heat transfer
Heat transfer, 58-60, 159-167
  by conduction, 161-162
  by convection, 162-163
  by radiation, 160-161
Heterocyclic nitrogen compounds, determination of, 107
Hexacyanoferrates, 119, 136-137
  determination of, 119, 124
Hexane, as solvent, 7, 108
Hexoses, 141
  monophosphates of, 141
Hydrocarbons, as solvent, 7, 99, 101, 108
  olefins in, 99
  total organoaluminum-reactable impurities in, 72
Hydrochloric acid, as titrant, 25, 62, 85, 88, 89, 94, 99, 113, 115
  determination in mixtures of acids, 86, 87, 98
  determination of, 39, 40, 67, 72
  heat of neutralization of, 78-80, 81-82
Hydrochlorothiazide, 95
Hydrofluoric acid, 87-88
Hydrogen bromide, as titrant for amines, 104
Hydrogen peroxide, as titrant, 132-133
  in "enthalpimetric indicator" reactions, 107, 152
Hydrogen orthophosphate, entropy titration of, 141
Hydrogen sulfate, entropy titration of, 141
Hydrogen sulfite, 133
Hydrolytic reactions, thermochemical studies of, 148
Hydroquinone, 130, 131
Hydroxy acids, determination of, 105
Hydroxybenzoic acids, determination of, 107

β-Hydroxyethyl-2-methyl-5-nitroimidazole, 106
5-Hydroxy-6-methyl-3,4-pyridine dimethanol-HCl, 106
8-Hydroxyquinoline-5-sulfonic acid, mixed ligand complexes, 150
Hydroxyl value, 70, 109
Hypochlorite, 62, 133
Hypophosphorous acid, 93
  in mixture of phosphoric acids, 93

Imides, determination of, 105, 107
Iminodiacetic acid, mixed ligand chelates, 150
Indium, organic acide chelate of, 151
Indole, 108
Initial slope method, basis of, 51-52
  in multistep reactions, 52, 142-144
  in single-step reactions, 42, 51, 103
  see also Thermal power, initial
Injection enthalpogram, 67; see also Direct injection enthalpimetry
Integral heat effect, 35, 41, 45, 47-53, 64, 65, 83, 116, 144-146
  definition of, 35
Iodide, as titrant, 113, 115, 133
  determination of, 62, 118, 124, 130, 131, 133
  metal complexes of, 136, 137, 147, 148
Iodine, as titrant, 54, 129, 133
  complexes of, 101
  determination of, 101, 133
Iodine value, 2, 108
Iodine monochloride, determination of olefinic unsaturation with, 108, 133
Ionic strength, influence on neutralization reaction, 82
Iron, analysis of, 69, 73, 74
  hexacyanoferrate complexes of, 136-137
Iron(II), as titrant, 62, 128, 129, 130, 132
  determination of, 62, 128, 130, 131, 132, 133
  organic acid chelate of, 151
Iron(III), as titrant, 131
  determination of, 149, 150
  fluoride complexes of, 149
  organic acid chelates of, 150, 151
  phenol complex of, 137
  phosphoric acid complexes of, 136
Isobutyl vinyl ether, as "enthalpimetric indicator" solvent, 107
Isooctane, as solvent, 141
Isopropanol, as solvent, 104, 107, 108
Isopropyl disulfide, 108
Isoquinoline, as titrant for organoaluminum compounds, 101

Keto-enols, 105
Ketones, as titrants for aluminum alkyls, 101
Kinetic effects, 5, 38-40, 43, 122-123
Kinetic titration, 5, 74-75
  differential method of, 75

Lactic acid dehydrogenase, 73
Lanthanides, basic nitrates of, 137
  complexes of, 148
  thermodynamic properties of, 148
  organic acid chelates of, 151
  see also Cerium; Lanthanum
Lanthanum, organic acid chelate of, 151
Lanthanum chloride, 115
Lattice energy, 114, 115
Lead, acetate complexes of, 136, 148
  as titrant, 119
  determination of, 69, 118, 119, 120, 123, 151
  hexacyanoferrates of, 119
  organic acid chelate of, 151
  reaction with fluoride, 149
Lead acetate, as titrant, 109
  titration with perchloric acid, 103
Lead nitrate, reaction with, alkali nitrates, 137
  alkali nitrites, 137
Lewis acids and bases, 7, 21, 40, 99-101, 135, 139-140
Limestone, analysis of, 122-123
Liquid mixtures, thermodynamic theory of, 66
Lithium, reaction with EDTA, 151
Lithium acetate, as titrant, 98
  titration with perchloric acid, 103
Lithium hydroxide, heat of neutralization, 79
LKB-Calorimeters, 182, 183, 198

Magnesite, analysis of, 69, 133
Magnesium, determination of, 3, 67, 123, 151
  alongside calcium, 3, 123, 152
  organic acid chelate of, 151
Magnesium acetate, titration with perchloric acid, 103
Maleic acid, 91, 98
Malonic acid, determination of, 91
  metal chelates of, 150
Manganese, as titrant, 128, 130
  determination of, 130, 133, 151
  in "enthalpimetric indicator" reactions, 107, 152
  organic acid chelate of, 151
Mercaptoacetic acid, 141
α-Mercaptocarboxylic acids, determination of, 131
Mercury(I), determination of, 122
Mercury(II), 2-aminoethanol complex of, 147
  cyanide complexes of, 148
  determination of, 118-119, 122, 124, 151
    alongside mercury(I), 122
  halide complexes of, 119, 147, 148
  organic acid chelate of, 151
Mercury acetate, titration with perchloric acid, 103
Mercury chloride, reactions of, 148
Mercury cyanide, reaction with thiourea, 147

INDEX

Mercury perchlorate, hydrolysis equilibria of, 148
Metal complexes, thermochemical studies of, 148
Metal-Cyanide coordination, thermodynamics of, 148
Metal-Halide coordination, thermodynamics of, 148
Metal ions, complexes with cyclic polyethers, 148
Metaphosphoric acid, determination of, 93
  in mixture of phosphoric acids, 93
Methanol, as solvent, 98, 99
Method of continuous variation, 138-139
  as entropy titration method, 139
m-Methoxybenzoic acid, 104
p-Methoxyphenol, 104
Methylamine, determination of, 88
  reaction with mercury chloride, 148
Methylcyclohexane, as solvent, 108
Methyl disulfide, 101
Methylene chloride, as solvent, 101, 104
α-Methyl styrene, as "enthalpimetric indicator" solvent, 107
Metol, 131
Microcalorimeters, 72, 196-198
Minerals, analysis of, 69, 119, 123, 149
Mixed ligand complexes, 148, 150
Molybdenum, determination of, 132, 133
  polyanions of, 95
Monochloroacetic acid, determination of, 84
  in mixture of acids, 86-87
  thermodynamic data for neutralization of, 83-84
Morpholine, as titrant, 99
  determination of, 88
  in mixtures with amines, 88
Mueller bridge, 172

Naphtols, 84, 85
Nickel, amino-acid complexes of, 148
  basic sulfates of, 136
  cobalt impurities in, determination of, 74
  determination of, 119, 150, 151, 152
  hexacyanoferrate compounds of, 119, 137
  organic acid chelates of, 151, 152
  pyrophosphate complexes of, 137
Nickel perchlorate, hydrolysis equilibria of, 148
Nicotine, 85, 87
Nitrilotriacetic acid, see NTA
o-Nitraniline, 108
Nitrates, basic, of lanthanides, 137
  lead complexes of, 137
  in nonaqueous solvent, determination of, 103
Nitrating acids, as titrants, 108
Nitric acid, as titrant, 94
  determination of, 72
  in ternary mixture, enthalpy of mixing, 66
  thermodynamic data for neutralization of, 79
Nitrite, as titrant, 69
  cadmium complexes of, 148
  lead complexes of, 137
4-Nitroacetanilide, 132
4-Nitroaniline, 132
Nitrobenzene, as solvent, 99
  determination of, 132
4-Nitrobenzoic acid, 132
Nitro compounds, determination of, 39, 40, 131, 132
Nitrogen compounds, organic, determination of, 85, 88, 95, 107
Nitrogen-containing bases, titration with tetraphenylborate, 121
Nitromethane, as solvent, 106
  in ternary mixture, enthalpy of mixing, 66
4-Nitrophenol, 104, 132
3-Nitrophthalic acid, 132
8-Nitroquinoline, 132
Nitroso compounds, determination of, 131, 132
4-Nitroso-N,N'-diethylaniline, 132
N-Nitrosodiphenylamine, 132
2-Nitroso-1-naphthol, 132
1-Nitroso-2-naphthol, 132
NTA, 107, 152

Octane, as solvent, 101
Oil, acetyl value of, 2, 109
Olefins, determination of, 7, 99, 108, 133
Organoaluminum compounds, determination of, 101
Organophosphorus pesticides, 73
Orthophosphates, as titrant, 123
  determination of, 103
Orthophosphoric acid, determination of, 25, 77, 91-93
  in mixture of phosphoric acids, 93
  iron(III)complexes of, 136
  thermodynamic data for neutralization of, 13, 91-93
Overflow method, 63
Oxalic acid, as titrant, 122-123
  determination of, 91, 130
  metal chelates of, 150
Oxine, 101

Palladium, cyanide complexes of, 148
  determination of, 124, 151
  organic acid chelate of, 151
Partial molar quantities, 8
Peltier effect, 167
Pentoses, 141
  monophosphates of, 141
Perchloric acid, as catalyst, 99
  as titrant, 95, 102-103, 107
  determination of, 80, 81, 122
  in "enthalpimetric indicator" reactions, 105
  thermodynamic data for neutralization of, 80, 81
Peroxodisulfate, see Persulfate
Persulfate, as titrant, 132
  determination of, 129, 130, 132
Petroleum products, acidity of, 2, 104
1,10-Phenanthroline, as titrant for

INDEX

organozinc compounds, 101
  metal complexes of, 147
Phenidone, 131
Phenol, determination of, 104, 131
  iron(III) complexes of, 137
  reaction with dimethylformamide, 13-14, 141
Phenolic acids, determination of, 105
Phenols, bromo-derivatives of, 104
  determination of, 84-85, 105, 107
Phenylalanat, 148
Phenylenediamines, 93
Phosphomolybdate, 74, 123
Phosphoric acid, *see* Orthophosphoric acid
Phosphorous acid, determination of, 93
  in mixtures of acids, 93
Phosphorus, derivatives of nitrogen compounds, 107
  determination of, 74, 123
Phthalic acid, 91
3-Picoline, 88
PID controller, 189
Pimelic acid, 91
Piperidine, 88
Plating solutions, analysis of, 69, 132
Platinum resistance thermometers, 169-170
  resistance ratio of, 170
Potassium, determination of, 119, 120, 122
Potassium acetate, titration with perchloric acid, 103
Potassium bromide, in eutectic melt $LiNO_3$-KNO_3, determination of, 109
  thermodynamic data, 109, 110
  see also Bromide
Potassium chloride, in eutectic melt $LiNO_3$-KNO_3, determination of, 109
  thermodynamic data, 109, 110
  reaction with $CdCl_2$ in KCl-$CdCl_2$ melt, 109
  see also Chloride
Potassium chromate, in eutectic melt $LiNO_3$-KNO_3, determination of, 109
  thermodynamic data, 109, 110
  reaction with perchloric acid, 95
Potassium hydrogen phthalate, 103, 106
  mixed ligand chelates, 150
Potassium hydroxide, as titrant, 104, 107
  thermodynamic data for neutralization of, 79
Potassium oxalate, reaction with thorium nitrate, 137
Precision thermostat, description of, 189-191
Precision titration calorimeter, description of, 185-191
Propionic acid, 84
Propyl disulfide, 101
Prussian blue, formation of, 148
Pseudoephedrine, 141
Purine bases, determination of, 107
Pyridine, as Lewis base, 100
  as solvent, 104
  derivatives, determination of, 107
  determination of, 85, 88
  in mixture of amines, 87, 88, 89, 104
  metal complexes of, 141, 147
Pyridine hydrochloride, 88
Pyrimidine, 141
  nucleotides of, 141
Pyrocatechol, mixed ligand chelates, 150
Pyrogallol, 107
Pyrrolidine, 88

Quinoline, 88
8-Quinolinol, as titrant for organozinc compounds, 101
Quinoxaline-2,3-dithiol, as titrant for nickel and selenium, 150

Resorcinol, determination of, 131
  in "enthalpimetric indicator" reaction, 107, 152

Salicyclic acid, mixed ligand complexes, 150
Sarcosine, 148
Section method, basis of, 43-46, 144
Selenium, determination of, 130, 150
Selenous acid, 130
Siemens-Martin slags, *see* Slags
Silicates, analysis of, 69, 123, 133
Silicon, determination of, 119, 149
Silver, as titrant, 62, 109, 118
  cyanide complexes of, 148
  determination of, 39, 62, 113, 115, 118, 119, 123, 124
  halide precipitation, thermodynamic data of, 114-115, 118
    concentration dependence of, 114-115
  hexacyanoferrate compounds of, 119, 137
  pyridine complexes of, 145-146
    thermodynamic data for complex formation, 145-146
  organic acid chelate of, 151
  thiosulfate complexes of, 137
Slags, analysis of, 69, 73, 119, 123, 133, 149
Sodium, determination of, 119
Sodium acetate, 103, 106
Sodium benzoate, 103, 106
Sodium carbonate, determination of, 94
  in mixture with hydroxide and sulfite, 94
  see also Carbonates
Sodium citrate, 103
Sodium formate, 106
Sodium hydrogen carbonate, 94
di-Sodium hydrogen orthophosphate, 25, 83, 93, 94
Sodium dihydrogen orthophosphate, 25, 93, 94
Sodium hydrogen sulfite, 133
Sodium hydroxide, as titrant, 3, 25, 39, 67, 72, 81, 82-94, 95, 120, 121
  determination of, 25
  in mixture with carbonate and sulfite, 94
  thermodynamic data for neutralization of, 78-80, 81, 82

influence of ionic strength on, 82
Sodium molybdate, 95
Sodium nitrite, as titrant, 69
   reaction with lead nitrate, 137
tri-Sodium orthophosphate, 25, 94
Sodium orthovanadate, 95
Sodium salicylate, 106
Sodium sulfate, 116
   sulfur content of, 124
Sodium sulfite, determination of, 131, 133
   in mixture with hydroxide and carbonate, 94
di-Sodium tetraborate, 94
Sodium tetraphenylborate, as titrant, 120-122
Sodium thiosulfate, 94, 129, 131
Sodium tungstate, 95
Specific heat, 30, 60
Steel, analysis of, 69, 73, 74, 119
Strontium, 122, 151
Succinic acid, 91, 107
Succinimide, 105
Sulfanilamide, 69
Sulfates, basic, 136
   determination of, 116-117, 123-124
   see also Sulfur, content
Sulfide, determination of, 124, 130
   in mixture with thiosulfate, 54
   see also Sulfur, content
Sulfites, 133
Sulfonal, sulfur content of, 124
Sulfonamides, 133
5-Sulfosalicyclic acid, mixed ligand complexes, 150
Sulfur, content, determination of, 73-74, 124, 130
   derivatives of nitrogen compounds, determination of, 107
   determination of, 73-74
Sulfuric acid, as titrant, 116-117, 118
   determination of, 13, 77, 89, 116-117
   in acetylating mixtures, 108-109
   in mixtures of acids, 86-87, 98
   in presence of hydrolyzable cations, 88
   in ternary mixtures, enthalpy of mixing, 66
   strength of, 70
   sulfur content in, 124
   thermodynamic data, 89, 116-117
Surfactants, analysis of, 2

Tangent method, basis of, 46-47, 142-144; *see also* Initial slope method
Tar products, 88
Tartaric acid, as chelating agent, 150
   determination of, 91
Temperature measurement, 56-58, 70, 167-176
TET, *see* Thermometric enthalpy titration
Tetra-n-butylammonium hydroxide, 107
Tetrahydrofuran, as Lewis base, 100, 139
Tetrahydropyran, as Lewis base, 100, 139
Tetraphenylborates, as titrants, 120-122
Tetrazoles, substituted, 141

Thallium, determination of, 120
   halogen complexes of, 148
Thallium(I)ethoxide, as titrant, 104
THAM, protonization of, 81
   as calibration standard, 81, 181
   thermodynamic data for, 81
Thermal conductivities, 162
Thermal power, balance of titration process, 33, 64
   caused by calorimeter, 33
   definition of, 29
   due to overflow, 63
   initial, 46-47, 48-49, 142-144, 145
   of dilution process, 31-32
   of heat transfer, 33
   of mixing, 30-31
   of reaction, 31-32, 48, 113-118, 142-144
   of stirring, 33
Thermistor bridge, 58, 172-176
   linearization of unbalance voltage of, 174-175
   recording of unbalance voltage of, 58, 175-176
Thermistors, 58, 70, 170-171
   linearization of temperature function of, 174-175
   stability of, 171
   temperature coefficient of, 170
Thermocouples, 167-169
   advantages of, 169
   criteria of usefullness, 167
   multijunction, 57, 168, 169
   properties of, 169
Thermogram, basic types of, 24-29, 67
   synonymous terms for, 24
   parts of, 28
   theoretical basis for analysis of, 24-54, 66-69, 139-140, 140-149
Thermometric enthalpy titration, definition of, 1
   description of, 56-58
   differential method, 58, 183-185
   equipment for, 56-58, 63-64, 181-193
   for determining enthalpy of mixing, 64-66
   in microanalysis, 58
   theoretical basis of, 24-43
   see also Thermogram
Thiocyanate, determination of, 118, 124
   in mixture with iodide, 118
   metal complexes of, 148, 149
Thiosulfate, determination of, 62, 129, 133
   metal complexes of, 137
Thiourea, determination of, 108, 131
   N-substituted derivatives, determination of, 131
Thorium, determination of, 151
   organic acid chelate of, 150, 151
   oxalate complexes of, 137
Thorium nitrate, 88
Threonine, reactions of, 148
Tin, determination of, 120, 128, 130, 149, 151
   fluoride complexes of, 149
   organic acid chelate of, 151

INDEX

Tin(IV)chloride, as Lewis acid, 7, 99-100, 139-140
  organotin(IV)chlorides, 100
Tiron, mixed ligand chelates, 150
Titanum, as titrant, 131
  determination of, 128, 130, 132
TITRA-THERMO-MAT, 191
Titration curve, see Thermogram
Toluene, as solvent, 141
  determination of, 108
p-Toluic acid, 104
o-Toluidine, 88
Trichloroacetic acid, determination of, 84
  thermodynamic data for neutralization of, 84
Triethylaluminum, as titrant, 99
  determination of, 101
Triethylamine, as titrant, 99, 101
  determination of, 106
1,3-Triethylammonium-2-propanol iodide, 106
Trifluoroethanol, 104
Trimethylamine, 88
Tris-(hydroxymethyl)-aminomethane, see THAM
TRONAC-Calorimeter, 183
Tungsten, determination of, 95
  reaction with hydrogen peroxide, 132
  polytungstates, 95
Tungstophosphoric acid, 95
Turnbull's blue, formation of, 148
Twin-cell titrator, 95, 183-185

Uranium, complexes of, 136, 137
  determination of, 130
  hydrolysis equilibria of salts of, 148
  reaction with hydrogen peroxide, 152
Uranyl sulfate, 88
Urea, 103, 108

Vanadium, determination of, 95, 132-133

Water, determination of, 70, 98-99

Yttrium, basic nitrate of, 137

Zinc, acetate complexes of, 148
  basic salts of, 136
  cyanide complexes of, 119, 148, 149
  determination of, 119, 120, 123, 149, 151, 152
  hexacyanoferrate compounds of, 119, 137
  organic acid chelates of, 151, 152
  1,10-phenantroline complexes of, 147
Zirconyl fluoride, 87-88

OHIO UNIVERSITY LIBRARY

Please return this book
finished with it.
be returned by